移动互联应用"赢在起点"系列图书

Android 移动开发项目化教程

主 编 李 扬
副主编 任学雯

教·学
资 源

航空工业出版社

北 京

内 容 提 要

本书共有 22 个项目，分为 5 个篇节：环境搭建篇、基础篇、进阶篇、数据处理篇和扩展篇。环境搭建篇介绍了 Android 系统的产生与发展过程、Android 开发环境搭建、Android 模拟器与开发终端以及 Android 工程创建与资源使用；基础篇介绍了 Android 常用界面控件、Android 常用界面布局、Android 单击事件的处理、Intent 实现消息传递以及 Activity 的生命周期；进阶篇介绍了高级用户界面设计、列表视图 ListView、自定义菜单、BroadCastReceiver 实现广播的接收与发送以及 Service 生命周期；数据处理篇介绍了 SharedPerference 与 XML、IO 操作与数据存储访问、SQLite 实现数据的存储与访问以及 ContentProvider 实现数据共享；扩展篇包括位置服务与百度地图实现地图定位、桌面小组件、NDK 的安装和使用以及 NDK 编译生成动态库。

本书涉及面广，涉及 Android 项目开发的各种知识，讲解明了清晰，可供 Android 初学者使用，也可作为各类院校的学生和相关培训班的学员的教材使用。

图书在版编目（CIP）数据

Android 移动开发项目化教程 / 李扬主编. -- 北京：航空工业出版社，2017.1（2024.2 重印）
ISBN 978-7-5165-1163-3

Ⅰ. ①A… Ⅱ. ①李… Ⅲ. ①移动终端－应用程序－程序设计－教材 Ⅳ. ①TN929.53

中国版本图书馆 CIP 数据核字(2017)第 013531 号

Android 移动开发项目化教程
Android Yidong Kaifa Xiangmuhua Jiaocheng

航空工业出版社出版发行
（北京市朝阳区京顺路 5 号曙光大厦 C 座四层　100028）
发行部电话：010-85672666　　010-85672683

捷鹰印刷（天津）有限公司印刷	全国各地新华书店经销
2017 年 1 月第 1 版	2024 年 2 月第 9 次印刷
开本：787×1092　1/16	字数：503 千字
印张：21.75	定价：58.00 元

随着 Android 系统的迅猛发展，它已经成为全球范围内具有广泛影响力的操作系统。Android 系统已经不仅仅是一款手机的操作系统，它越来越广泛地被应用于平板电脑、可佩戴设备、电视、数码相机等设备上，这也造就了目前 Android 开发人才需求的快速增长，从大趋势上看，Android 软件人才的需求将越来越大。

本书主要围绕 Android 开发技术，循序渐进地讲解了开发 Android 应用所需要的基础知识，其中包括开发环境、基本组件应用、事件处理、数据存储以及 GPS 定位等。

本书特色

一本好教材，应该易教、易学，让学生轻松学到实用的知识；一本好教材，应该内容安排合理、体例新颖、实用；一本好教材，应该概念准确，语言精炼，讲解通俗易懂；一本好教材，应该图文并茂，案例丰富、典型、实用。

具体来说，本书具有以下几个特点：

➢ **知识点融入案例**：本书摒弃枯燥的理论，没有长篇的对于某个知识点的解释，以及属性、方法的大量罗列，这些东西读者通过官方的帮助文档都可以获得。文中知识点都是通过具体案例来进行讲解，便于理解和掌握。

➢ **结构合理，易学易用**：从用户的实际需要出发，由浅入深讲解所需知识。读者既可以按照本书编排的章节顺序进行学习，也可以根据自己的需求对某一章节进行针对性的学习。

➢ **精心安排内容，符合岗位需要**：本书精心挑选与实际应用紧密相关的知识点和案例，从而让读者在学完本书后，能马上在实践中应用学到的技能。

本书主要内容

本书共分 5 个篇节，具体内容如下：

➢ **环境搭建篇**：介绍了 Android 系统的产生与发展过程、Android 开发环境搭建、Android 模拟器与开发终端以及 Android 工程创建与资源使用；

➢ **基础篇**：介绍了 Android 常用界面控件、Android 常用界面布局、Android 单击事件的处理、Intent 实现消息传递以及 Activity 的生命周期；

➢ **进阶篇**：介绍了高级用户界面设计、列表视图 ListView、自定义菜单、BroadCastReceiver 实现广播的接收与发送以及 Service 生命周期；

➢ **数据处理篇：** 介绍了 SharedPerference 与 XML、IO 操作与数据存储访问、SQLite 实现数据的存储与访问以及 ContentProvider 实现数据共享；

➢ **扩展篇：** 包括位置服务与百度地图实现地图定位、桌面小组件、NDK 的安装和使用以及 NDK 编译生成动态库。

本书配套资源

本书配有精美的教学课件，并且书中涉及的程序代码都已整理并打包，读者可以登录文旌综合教育平台"文旌课堂"（www.wenjingketang.com）下载。读者如果在学习过程中遇到什么问题，也可通过本网站获得帮助。

为学习贯彻党的二十大精神，提升课程铸魂育人效果，本书专门在扉页"教•学资源"二维码中设计了相应栏目，以引导学生践行社会主义核心价值观，涵养学生奋斗精神、敬业精神、奉献精神、创新精神、工匠精神、法制精神、绿色环保意识等。

本书由天津现代职业技术学院李扬博士任主编，任学雯老师任副主编。同时，魏宁、贾珺、宗哲玲老师也参与了本书的编写。在编写过程中，作者参考了大量国内外出版的相关教材和资料，在此谨向相关作者致以诚挚的谢意。由于水平有限，书中难免存在疏漏之处，敬请广大读者批评指正。

环境搭建篇

项目一　Android 系统概述 ·················· 2
一、项目要求 ·································· 2
二、项目相关知识 ···························· 2
三、项目实施过程 ···························· 3
 1. Android 系统的产生与发展 ········ 3
 2. Android 系统的优势 ················· 4
 3. Android 平台架构 ···················· 4
 4. Android 平台特性 ···················· 7
 5. Android 市场 ·························· 7
四、项目思考与扩展 ······················· 10

项目二　Android 开发环境设置 ········ 11
一、项目要求 ································ 11
二、项目相关知识 ·························· 11
 1. 操作系统要求 ······················· 11
 2. JDK（Java Development Kit）······ 12
 3. Eclipse ································ 12
 4. Android SDK ························ 12
 5. ADT（Android Development Tools）································ 12

 6. Android Studio 集成开发工具 ······ 13
三、项目实施过程 ·························· 13
 1. 安装 JDK ····························· 13
 2. 安装 Eclipse ························· 19
 3. 安装 SDK ····························· 21
 4. 安装 ADT ···························· 23
 5. Android Studio 的下载、安装和使用 ··································· 26
四、项目思考与扩展 ······················· 32

项目三　Android 模拟器与开发终端 ···· 33
一、项目要求 ································ 33
二、项目相关知识 ·························· 33
 1. Android 模拟器简介 ················ 33
 2. AVD 和真实设备的区别 ············ 34
三、项目实施过程 ·························· 35
 1. 创建 Android 虚拟设备 ············· 35
 2. 连接真实 Android 设备 ············· 37
四、项目思考与扩展 ······················· 38

项目四　Android 工程创建与资源使用 ……39
一、项目要求 ……39
二、项目相关知识 ……39
1. Android SDK 软件开发包 ……39
2. Android 程序结构 ……40
3. DDMS ……41
三、项目实施过程 ……42
1. 建立 Android 项目 ……42
2. DDMS 的使用 ……46
四、项目思考与扩展 ……49

基础篇

项目五　Android 常用界面控件 ……52
一、项目要求 ……52
二、项目相关知识 ……52
1. Android 视图类 ……52
2. 本文框 TextView ……53
3. 按钮 ……54
4. 单选按钮和复选框按钮 ……55
5. 列表选择框 ……56
三、项目实施过程 ……56
1. 项目创建 ……57
2. XML 布局文件的开发 ……57
四、项目思考与扩展 ……63

项目六　Android 常用界面布局 ……64
一、项目要求 ……64
二、项目相关知识 ……64
1. 线性布局 ……65
2. 表格布局 TableLayout ……67
3. 帧布局 ……71
4. 相对布局 RelativeLayout ……73
三、项目实施过程 ……76
1. 创建工程 ……77
2. XML 布局文件的开发 ……77

四、项目思考与扩展 ……80

项目七　Android 单击事件的处理 ……81
一、项目要求 ……81
二、项目相关知识 ……81
1. 事件监听原理 ……82
2. findViewById()方法 ……83
3. 按钮单击事件的四种方法 ……83
三、项目实施过程 ……86
1. 创建工程 ……86
2. XML 布局文件的开发 ……87
3. Java 文件的开发 ……88
四、项目思考与扩展 ……91

项目八　Intent 实现消息传递 ……92
一、项目要求 ……92
二、项目相关知识 ……92
1. Intent 概述 ……92
2. 显性 Intent ……93
3. 显性 Intent 的数据传递 ……96
4. 隐性 Intent ……97
三、项目实施过程 ……99
1. 创建工程 ……99
2. XML 布局文件的开发 ……100
3. Java 文件的开发 ……111

四、项目思考与扩展 …………… 113

项目九　Activity 的生命周期………… 115
　一、项目要求 …………………… 115
　二、项目相关知识 ……………… 116
　　1．Android 生命周期与进程优先级 …… 116
　　2．Activity 的生命周期 …………… 116
　　3．Log 类的使用 …………………… 118
　三、项目实施过程 ……………… 118
　四、项目思考与扩展 …………… 129

进阶篇

项目十　高级用户界面设计 …………… 132
　一、项目要求 …………………… 132
　二、项目相关知识 ……………… 133
　　1．自动完成文本框
　　　AutoCompleteTextView ………… 133
　　2．进度条 ProgressBar ……………… 136
　　3．拖动条 SeekBar ………………… 140
　　4．星级评分条 RatingBar ………… 142
　　5．选项卡 TabHost ………………… 144
　　6．图像切换器 ImageSwitcher …… 148
　　7．画廊视图 Gallery ……………… 152
　　8．消息提示框 Toast ……………… 156
　三、项目实施过程 ……………… 157
　　1．创建工程 ……………………… 157
　　2．XML 布局文件的开发 ………… 158
　　3．Java 文件的开发 ……………… 159
　四、项目思考与扩展 …………… 162

项目十一　列表视图 ListView ………… 163
　一、项目要求 …………………… 163
　二、项目相关知识 ……………… 163
　　1．ListView ………………………… 163
　　2．直接使用 ListView 组件
　　　创建 ListView …………………… 164
　　3．让 Activity 继承 ListActivity
　　　实现列表 ………………………… 166
　三、项目实施过程 ……………… 168
　　1．创建工程 ……………………… 168
　　2．XML 布局文件的开发 ………… 169
　　3．Java 文件的开发 ……………… 170
　四、项目思考与扩展 …………… 171

项目十二　自定义菜单 ………… 172
　一、项目要求 …………………… 172
　二、项目相关知识 ……………… 172
　三、项目实施过程 ……………… 173
　　1．创建工程 ……………………… 173
　　2．XML 布局文件的开发 ………… 173
　　3．Java 文件的开发 ……………… 178
　四、项目思考与扩展 …………… 181

**项目十三　BroadCastReceiver 实现
　　　　　广播的接收与发送** ……… 183
　一、项目要求 …………………… 183
　二、项目相关知识 ……………… 183
　　1．广播 …………………………… 183
　　2．静态和注册广播接收器 ………… 184
　三、项目实施过程 ……………… 184
　　1．创建工程 ……………………… 184

2．XML 布局文件的开发…………185
3．Java 文件的开发………………188
四、项目思考与扩展………………192

项目十四　Service 生命周期…………194
一、项目要求………………………194
二、项目相关知识…………………194
　　1．Service 简介…………………194
　　2．Service 生命周期……………195

3．跨进程服务简介………………196
三、项目实施过程…………………196
　　1．工程创建……………………196
　　2．XML 布局文件的开发………197
　　3．Java 文件的开发……………199
　　4．扩展练习……………………203
　　5．项目验证……………………203
四、项目思考与扩展………………205

数据处理篇

项目十五　SharedPerference 与 XML……208
一、项目要求………………………208
二、项目相关知识…………………209
　　1．SharedPerference 简介………209
　　2．SAX 的简介…………………209
三、项目实施过程…………………210
　　1．创建工程……………………210
　　2．XML 布局文件的开发………210
　　3．Java 程序的开发……………214
　　4．扩展练习……………………219
　　5．项目验证……………………221
四、项目思考与扩展………………223

项目十六　IO 操作与数据存储访问……224
一、项目要求………………………224
二、项目相关知识…………………224
　　1．内部文件存储………………224
　　2．外部文件存储………………225
　　3．资源文件使用………………225
三、项目实施过程…………………226
　　1．创建工程……………………226

2．XML 布局文件的开发………226
3．Java 文件的开发……………229
4．扩展练习……………………233
5．项目验证……………………234
四、项目思考与扩展………………236

项目十七　SQLite 实现数据的存储与访问………………………237
一、项目要求………………………237
二、项目相关知识…………………238
　　1．SQLite 简介…………………238
　　2．SQLite 对数据库的操作……239
　　3．动态广播的使用……………241
三、项目实施过程…………………242
　　1．工程创建……………………242
　　2．XML 布局文件的开发………242
　　3．Java 文件的开发……………250
　　4．扩展练习……………………264
　　5．项目验证……………………265
四、项目思考与扩展………………267

项目十八　ContentProvider 实现数据共享 ………… 268
一、项目要求 ……………………… 268
二、项目相关知识 ………………… 268
　1. ContentProvider 介绍 ………… 268
　2. 系统通讯录核心操作代码 …… 270
三、项目实施过程 ………………… 272
　1. 工程创建 …………………… 272
　2. XML 布局文件的开发 ……… 273
　3. Java 文件的开发 …………… 277
　4. 项目验证 …………………… 286
四、项目思考与扩展 ……………… 288

扩展篇

项目十九　位置服务与百度地图实现地图定位 ………… 290
一、项目要求 ……………………… 290
二、项目相关知识 ………………… 290
　1. 基于位置的服务简介 ………… 290
　2. LocationManager 的基本用法 … 291
　3. 获取 GPS 定位信息的步骤 … 293
　4. 使用 MapView 显示定位的过程 … 293
三、项目实施过程 ………………… 294
　1. 工程创建 …………………… 294
　2. XML 布局文件的开发 ……… 296
　3. Java 文件的开发 …………… 300
　4. 项目验证 …………………… 305
四、项目思考与扩展 ……………… 308

项目二十　桌面小组件 ………… 309
一、项目要求 ……………………… 309
二、项目相关知识 ………………… 309
三、项目实施过程 ………………… 310
　1. 创建工程 …………………… 310
　2. XML 布局文件的开发 ……… 311
　3. Java 文件的开发 …………… 314
　4. 项目验证 …………………… 318

四、项目思考与扩展 ……………… 319

项目二十一　NDK 的安装和使用 …… 320
一、项目要求 ……………………… 320
二、项目相关知识 ………………… 320
　1. NDK（Native Development Kit）… 320
　2. Cygwin ……………………… 321
　3. so 文件 ……………………… 321
三、项目实施过程 ………………… 321
　1. Cygwin ……………………… 321
　2. NDK 环境参数的设置 ……… 323
　3. Windows 环境下利用 NDK 生成 SO ……………………… 324
四、项目思考与扩展 ……………… 325

项目二十二　NDK 编译生成动态库 …… 326
一、项目要求 ……………………… 326
二、项目相关知识 ………………… 326
三、项目实施过程 ………………… 327
　1. 创建工程 …………………… 327
　2. Java 文件的开发 …………… 328
四、项目思考与扩展 ……………… 334

参考文献 ……………………… 335

环境搭建篇

项目一 Android 系统概述

【本章导读】

　　Android 系统就是一个开放式的移动互联网操作系统，目前已经成为应用最广的移动互联网平台。Android 最早由 Andy Rubin 创办，于 2005 年被 Google 收购。2007 年 11 月 5 日，Google 正式发布 Android 平台。在 2010 年底，Android 已经超越称霸 10 年的诺基亚 Symbian 系统，成为全球最受欢迎的智能手机平台。HTC、Samsung、Motorola、LG、Sony Ericsson 等手机厂商均采用 Android 平台。本章，我们就来先了解一下什么是 Android 操作系统。

一、项目要求

1. 了解 Android 平台的历史、现状。
2. 掌握 Android 平台架构及特性。
3. 了解 Android 应用的市场。

二、项目相关知识

　　Java 语言作为服务器端编程语言，Java EE 平台发展得非常成熟，而且一直是电信、移动、银行、证券、电子商务应用的首选平台。但在客户端应用开发方面，Java 语言一直表现不佳。Android 系统的出现改变了这种局面，它目前已经成为应用最广的智能手机、平板电脑操作系统，采用 Java 语言开发的 Android 应用程序也越来越多。

　　不过需要指出的是，Android 平台能够运行的硬件只是智能手机、平板电脑等便携式设备，这些设备的计算能力、数据存储能力都是有限的，因此不宜在 Android 平台上部署大型企业级应用。但 Android 应用可以纯粹客户端应用的角色出现，然后通过网络与传统大型应用交互，充当大型企业应用的网络客户端，比如现在已经出现的淘宝 Android 客户端、赶集网 Android 客户端，它们都是这种发展趋势下的产物。

三、项目实施过程

1. Android 系统的产生与发展

Android 是由 Andy Rubin 创立的一个手机操作系统，后来被 Google 收购。Google 希望与各方共同建立一个标准化、开放式的移动电话软件平台，从而在移动产业内形成一个开放式的操作平台。

Google 于 2007 年 11 月 5 日发布了 Android 1.0 手机操作系统，这个版本的 Android 系统还没有赢得广泛的市场支持。

2009 年 5 月，Google 发布了 Android 1.5，该版本的 Android 提供了一个非常"豪华"的用户界面，而且提供了蓝牙连接支持。这个版本的 Android 吸引了大量开发者的目光。

接下来，Android 的版本更新得较快，随着新版本的不断发布，Android 的功能也日益强大，涌现了很多流行的应用程序，也催生了一大批优秀组件。

当前市场上常见的手机操作系统有：

- **iOS**：Apple 公司的手机、平板操作系统，市场占有率较高。
- **Windows Phone**：Microsoft 公司的手机操作系统，2012 年发布的最新版本为 Windows Phone 8，但局势依然不够明朗。2015 年，微软推送 Windows 10 移动版，在 Windows Phone 8 的基础上改进了一些功能的操作方式。
- **Symbian**：2012 年诺基亚彻底放弃开发塞班系统，Symbian 已被放弃、基本被淘汰。
- **BlackBerry**：即将被淘汰。

目前 Android 系统的市场占有率已经远超 iOS，而 Windows Phone 作为 Microsoft 最后的"赌注"，自然也是全力以赴，希望至少能与 iOS、Android 三足鼎立，但目前局势似乎并不乐观。无论从哪个角度来看，Android 已经成为最主流的手机操作系统。

就目前国内环境来说，已有大量手机厂商开始生产 Android 操作系统的手机，因为 Android 手机平台是一个真正开放式的平台，无需支付任何费用即可使用。出于节省研发费用的考虑，不管是对于知名手机生产厂商，还是大量的山寨手机厂商，Android 操作平台都是一个不错的选择。

据 2015 年新数据分析,全球智能手机操作系统市场份额中,Android 操作系统的占有率达 80%以上。目前,已发布搭载 Android 系统的手机厂商包括:摩托罗拉、三星、HTC、索尼爱立信、LG 等,国内厂商如华为、联想、中兴等也都开始发布搭载 Android 系统的手机。

2. Android 系统的优势

为什么安卓能在这么多的智能系统中脱颖而出,成为市场占有率第一的手机系统呢?要想分析其原因,需要先了解它的巨大优势,分析究竟是哪些优点吸引了厂商和消费者的青睐。

> 开源

这是 Android 能够快速成长的最关键因素。在 Android 之前,没有任何一个智能操作系统的开源程度能够像 Android 一样。Android 的开源,打破以往操作系统平台的授权模式,不但降低了厂商的成本也赋予了他们更多自由发挥的空间,更提升了他们支持 Android 的热情,这是 Android 平台能够快速成熟、快速成长的源泉。

> 联盟

联盟战略是 Android 能够攻城拔寨的另一大法宝。谷歌为 Android 成立的开放手机联盟(OHA)不但有摩托罗拉、三星、HTC、索尼爱立信等众多大牌手机厂商拥护,还受到了手机芯片厂商和移动运营商的支持,仅创始成员就达到 34 家。

开源、联盟,Android 凝聚了几乎遍布全球的力量,这是 Android 形象及声音能够被传到全球移动互联网市场每一个角落的根本原因。

> 技术

Android 系统的底层操作系统是 Linux,Linux 作为一款免费、易得、可以任意修改源代码的操作系统,吸收了全球无数程序员的精华。另外,Linux 作为一种嵌入式操作系统,使得 Android 能够很方便地被应用、移植到各种平台并快速发展。同时,Android 平台较快的版本更新速度使得手机硬件性能不断向最优方向发展,也使 Android 可玩、好玩、容易玩的特征越来越显著。

3. Android 平台架构

Android 系统的底层建立在 Linux 系统之上,该平台由操作系统、中间件、用户界面和应用软件 4 层组成,如图 1-1 所示。它采用一种被称为软件叠层(Software Stack)的方式进行构建。这种软件叠层结构使得层与层之间相互分离,明确各层的分工。这种分工保证了层与层之间的低耦合,当下层的层内或层下发生改变时,上层应用程序无需任何改变。

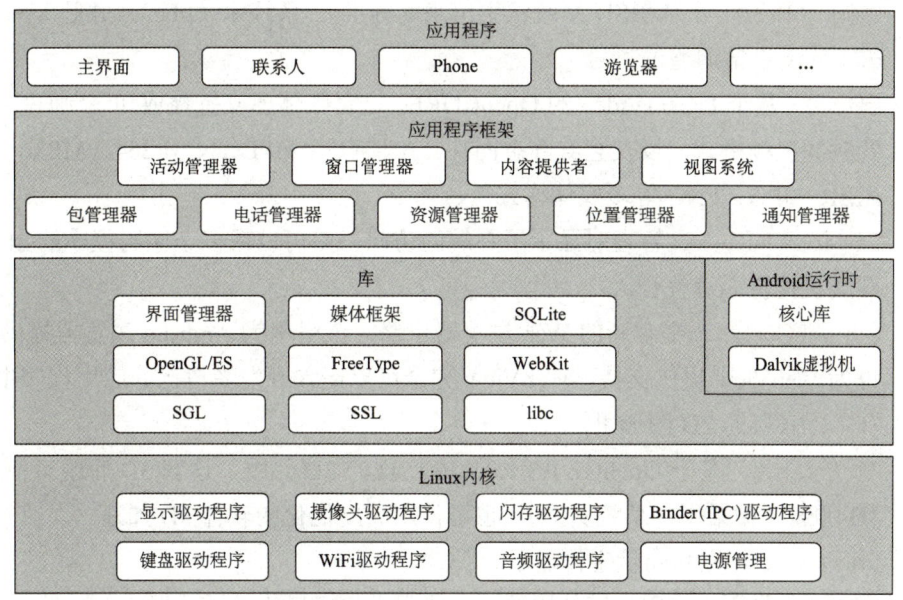

图 1-1 显示了 Android 系统的体系结构

1）应用程序层

Android 系统将会包含系列的核心应用程序，包括电子邮件客户端、SMS 程序、日历、地图、浏览器、联系人等。这些应用程序都是用 Java 编写的。这也是本书所介绍的主要内容——编写 Android 系统上的应用程序。

2）应用程序框架

前面已经提到，本书所要介绍的内容就是开发 Android 应用程序，当我们开发 Android 应用程序时，就是面向底层的应用程序框架进行的。从这个意义上来看，Android 系统上的应用程序是完全平等的，不管是 Android 系统提供的程序，还是普通开发者提供的程序，都可以访问 Android 提供的 API 框架。

Android 应用程序框架提供了大量 API 供开发者使用，关于这些 API 的具体功能和用法则是本书后面要详细介绍的内容，此处不再展开阐述。

应用程序框架除可作为应用程序开发的基础之外，也是软件复用的重要手段，任何一个应用程序都可发布它的功能模块——只要发布时遵守了框架的约定，那么其他应用程序也可使用这个功能模块。

3）函数库与运行时

Android 包含一套被不同组件所使用的 C/C++库的集合。一般来说，Android 应用开发者不能直接调用这套 C/C++库集，但可以通过它上面的应用程序框架来调用这些库。

下面列出一些核心库：

➢ **系统 C 库**：一个从 BSD 系统派生出来的标准 C 系统库（libc），并肚专门为嵌入式 Linux 设备调整过。
➢ **媒体库**：基于 PacketVideo 的 OpenCORE，这套媒体库支持播放和录制许多流行的音频和视频格式，以及查看静态图片。主要包括 MPEG4、H.264、MP3、AAC、AMR、JPG、PNG 等多媒体格式。
➢ **Surface Manager**：管理对显示子系统的访问，并可以对多个应用程序的 2D 和 3D 图层机提供无缝整合。
➢ **LibWebCore**：一个全新的 Web 浏览器引擎，该引擎为 Android 浏览器提供支持，也为 WebView 提供支持，WebView 完全可以嵌入开发者自己的应用程序中。
➢ **SGL**：底层的 2D 图形引擎。
➢ **3D libraries**：基于 OpenGL ES 1.0 API 实现的 3D 系统，这套 3D 库既可使用硬件 3D 加速（如果硬件系统支持），也可使用高度优化的软件 3D 加速。
➢ **FreeType**：位图和向量字体显示。
➢ **SQLite**：供所有应用程序使用的、功能强大的轻量级关系数据库。

Android 运行时由两部分组成：Android 核心库集和 Dalvik 虚拟机。其中核心库集提供了 Java 语言核心库所能使用的绝大部分功能，而虚拟机则负责运行 Android 应用程序。

每个 Android 应用程序都运行在单独的 Dalvik 虚拟机内（即每个 Android 应用程序对应一条 Davlik 进程），Dalvik 专门针对同时高效地运行多个虚拟机进行了优化，因此 Android 系统以方便的实现对应用程序进行隔离。

由于 Android 应用程序的编程语言是 Java，因此有些人会把 Dalvik 虚拟机和 JVM 搞混，但实际上二者存在区别：Dalvik 并未完全遵守 JVM 规范，两者也不兼容。实际上，JVM 虚拟机运行的是 Java 字节码（通常就是.class 文件），但 Dalvik 运行的是其专有的 dex（Dalvik Executable）文件。JVM 直接从.class 文件或 JAR 包中加载字节码然后运行；而 Dalvik 则无法直接从.class 文件或 JAR 包中加载字节码，它需要通过 DX 工具将应用程序的所有.class 文件编译成.dex 文件，Dalvik 则运行该.dex 文件。

Dalvik 虚拟机非常适合在移动终端上使用，相对于在 PC 或服务器上运行的虚拟机而言，Dalvik 虚拟机不需要很快的 CPU 计算速度和大量的内存空间，它主要有如下两个特点。

➢ **运行专有的.dex 文件**。专有的.dex 文件减少了.class 文件中的冗余信息，而且会把所有.class 文件整合到一个文件中，从而提高运行性能；而且 DX 工具还会对.dex 文件进行一些性能的优化。
➢ **基于寄存器实现**。大多数虚拟机（包括 JVM）都是基于栈的，而 Dalvik 虚拟机则是基于寄存器的。一般来说，基于寄存器的虚拟机具有更好的性能表现，但在硬件通用性上略差。

Dalvik 虚拟机依赖于 Linux 内核提供的核心功能，如线程和底层内存管理。

4）Linux 内核

Android 系统建立在 Linux 2.6 之上。Linux 内核提供了安全性、内存管理、进程管理、网络协议栈和驱动模型等核心系统服务。除此之外，Linux 内核也是系统硬件和软件叠层之间的抽象层。

4. Android 平台特性

Android 平台具有如下特性：
- 允许重用和替换组件的应用程序框架；
- 专门为移动设备优化的 Dalvik 虚拟机；
- 基于开源引擎 WebKit 的内置浏览器；
- 自定义的 2D 图形库提供了最佳的图形效果，此外还支持基于 OpenGL ES 1.0 规范的 3D 效果（需要硬件支持）；
- 支持数据结构化存储的 SQLite；
- 支持常见的音频、视频和图片格式（例如 MPEG4、H.264、MP3、AAC、AMR、JPG、PNG、GIF）；
- GSM 电话（需要硬件支持）；
- 蓝牙、EDGE、3G 和 WiFi（需要硬件支持）；
- 摄像头、GPS、指南针和加速计（需要硬件支持）；
- 包括设备模拟器、调试工具、优化工具和 Eclipse 开发插件等丰富的开发环境。

5. Android 市场

Google 公司为 Android 平台提供的在线应用商店，Android 平台用户可以在该市场中浏览、下载和购买第三方人员开发的应用程序。对于开发人员，有两种挣钱的方式：第一种方式是卖软件，开发人员可以获得该应用售价的 70%，其余 30%作为其他费用；第二种方式是加广告，将自己的软件定为免费软件，通过增加广告链接，靠单击率挣钱。

因为 Android 系统的免费和开源，也因为系统本身强大的功能性，使得 Android 系统不仅被用于手机设备上，而且也被广泛用于其他职能设备中。下面介绍几款常见的搭载 Android 系统的智能设备。

1）Android 智能电视

Android 智能电视，顾名思义是搭载了安卓操作系统的电视，使得电视智能化，能让电视机实现网页浏览、视频电影观看、聊天办公游戏等，与平板电脑和智能手机一样的功能。其凭借安卓系统让电视实现智能化的提升，数十万款安卓市场的应用、游戏等内容随意安装。例如海尔的 MOOKA 模卡 U42H7030 便是一款搭载 Android 4.2 系统的智能电视，如图 1-2 所示。

图 1-2　搭载 Android 4.2 系统的智能电视

2）Android 机顶盒

Android 机顶盒是指像智能手机一样，具有全开放式平台，搭载了安卓操作系统，可以由用户自行安装和卸载软件、游戏等第三方服务商提供的程序，通过此类程序来不断对电视的功能进行扩充，并可以通过网线、无线网络来实现上网冲浪的新一代机顶盒总称。

Android 机顶盒不仅仅是一个高清播放器，更具有一种全新的人机交互模式，既区别于电脑，又有别于触摸屏，通过使用 Android 机顶盒，可以让电视具有上网、看网络视频、玩游戏、看电子书、听音乐等功能。Android 机顶盒配备红外感应条，遥控器一般采用空中飞鼠，这样就可以方便地实现触摸屏上的各种单点操作，可以方便地在电视上玩愤怒的小鸟、植物大战僵尸等经典游戏。乐视公司的 LeTV 机顶盒便是基于 Android 打造的，如图 1-3 所示。

图 1-3　基于 Android 的 LeTV 机顶盒

3）智能手表

智能手表是将手表内置智能化系统、搭载智能手机系统，连接于网络可实现多功能，如同步手机中的电话、短信、邮件、照片、音乐等。LG 公司采用谷歌 Android Wear 操作系统开发了一款名为 G Watch 的智能手表，该产品于 2014 年第二季度发布，如图 1-4 所示。

图 1-4 搭载 Android Wear 系统的 G Watch

4）游戏机

Android 游戏机就像 Android 智能手表一样，在 2013 年出现了爆炸式增长。在 CES 展会上，NVIDIA 的 Project Shield 掌上游戏主机以绝对震撼的姿态亮相，之后又有 Ouya 和 Gamestick 相继推出。不久前，Mad Catz 也发布了一款 Android 游戏机。

5）智能家居

智能家居是以住宅为平台，利用综合布线技术、网络通信技术、安全防范技术、自动控制技术、音视频技术将家居生活有关的设施集成，构建高效的住宅设施与家庭日常事务的管理系统，提升家居安全性、便利性、舒适性、艺术性，并实现环保节能的居住环境。

智能家居是在互联网影响之下的物联化体现。乐得威公司的 GW-9311 智能主机产品便是一款 Android 智能家居产品，如图 1-5 所示。

图 1-5 乐得威公司的 GW-9311 智能主机

上述智能设备只是冰山一角，随着物联网和云服务的普及和发展，将有更多的智能设备诞生。到那个时候，Android 系统更是翻云覆雨，将拥有一个更美好的未来。

四、项目思考与扩展

通过 Android 市场平台,下载并安装 Android 软件。

项目二　Android 开发环境设置

【本章导读】

古人云：工欲善其事，必先利其器。在进行 Android 移动终端的开发和设计之前，必须先要搭建一个易于开发的环境。本章项目，我们就来学习 Android 开发环境的下载与搭建。

一、项目要求

1. 掌握 Android 应用开发环境的搭建过程。
2. 掌握 JDK、Android SDK、ADT 的安装与环境变量的配置。
3. 熟悉 Android 编程工具 Eclipse 的安装使用。
4. 熟悉 Android 集成开发工具 Android Studio 的使用。

二、项目相关知识

Android 系统开发可以区分成两大部分：一是 Android 系统的应用程序开发，二是 Android 系统本身架构的开发。Android 应用程序的开发着重在如何应用 Framework 框架及 Android SDK，其主要使用的编程语言是 Java 语言。

1. 操作系统要求

在搭建 Android 的开发环境之前，首先要了解 Android 对 PC 操作系统的要求。从 Android 官方网站上可以看到，Android 支持在 Windows、Mac OS、Linux 等操作系统上进行程序开发。Android SDK 对操作系统的要求如表 2-1 所示。

表 2-1　Android SDK 对操作系统的要求

操作系统	要　　求
Windows	Windows XP（32 位）
	Visa（32 或 64 位）
	Windows 7（32 或 64 位）
Mac OS	10.5.8 或更新（仅支持 x86）
Linux	需要 GNU C Library（glibc）2.7 或更新，在 Ubuntu 系统上需要 8.04 版或更新，64 位版本必须支持 32 位应用程序

2. JDK（Java Development Kit）

Android 系统是基于 Java 语言的，所以在开发 Android 程序时，应首先保证机器具备 Java 环境，Java 的编程环境就是 JDK。JDK（Java Development Kit）是整个 Java 的核心，包括 Java 运行环境、Java 工具和 Java 基础的类库。

3. Eclipse

Eclipse 是一个非常强大并具有集成开发环境的开发工具，可以支持 Java、C、C++等多种语言。由于 Android 的开发需要使用 Java 语言，因此开发人员需要下载集成 Java 版本的 Eclipse 开发环境。目前，官网提供了 3 个版本，分别如下：

（1）Eclipse IDE for Java Developers

（2）Eclipse IDE for Java EE Developers

（3）Eclipse Classic

在实际开发中，我们只需选择其中的一种进行下载即可。

4. Android SDK

SDK（Software Development Kit，软件开发工具包）是用于为特定的软件包、软件框架、硬件平台、操作系统等建立应用软件的开发工具的集合。因此，Android SDK 指的是 Android 专属的软件开发工具包，该开发包为我们提供了在 Android 开发时所需要的类库及相应方法。进行 Android 移动应用程序开发之前，需要安装 Android SDK 开发包，开发人员可以在 Google 提供的官方网站上下载 Android SDK 开发包。

5. ADT（Android Development Tools）

安装 JDK、Eclipse 和 Android SDK 之后，便具备了 Android 开发所需要的最基本条件，但是这样的开发环境还不够人性化。为了使开发变得便捷，我们可以使用 Google 提供的

Eclipse 插件 ADT（Android Development Tools）。在 Eclipse 编译 IDE 环境中，安装 ADT，可以为 Android 开发提供开发工具的升级或者变更。

ADT 插件扩展了 Eclipse 的许多功能，例如：Android 平台的 API、使用 SDK 来调试 Android 程序、进行 APK 文件的导出等。安装 ADT 插件有两种方式：一种是通过 Eclipse 程序在线进行更新安装；另一种自行下载 ADT 手动进行配置。

6. Android Studio 集成开发工具

除了前面的开发环境，我们还可以使用 Android Studio，它是一个基于 IntelliJ IDEA 的 Android 集成开发工具。Android Studio 是 Google 于 2013 年 I/O 大会针对 Android 开发推出的新的开发工具，目前很多开源项目都已经在采用。

在 IDEA 的基础上，Android Studio 提供：

（1）基于 Gradle 的构建支持；
（2）Android 专属的重构和快速修复；
（3）提示工具以捕获性能、可用性、版本兼容性等问题；
（4）支持 ProGuard 和应用签名；
（5）基于模板的向导来生成常用的 Android 应用设计和组件；
（6）功能强大的布局编辑器，可以让你拖拉 UI 控件并进行效果预览。

三、项目实施过程

1. 安装 JDK

（1）登录 Oracle 官方网站，找到 Downloads 下面 Java SE 的下载界面，具体网址为 http://www.oracle.com/technetwork/Java/Javase/downloads/index.html，如图 2-1 所示。

（2）单击 JDK 下面对应的 DOWNLOAD 按钮，出现如图 2-2 所示的下载页面，这里有不同版本的 JDK 可供下载。根据自己电脑的配置选择符合的 JDK，笔者在此选择的是 Windows X86 版本。

> JDK 版本格式 JDK XuYY 的含义：
> 其中 X 代表版本号，例如 JDK7、JDK8；u 代表 update 首字母；YY 代表小版本号一般都是对当前大版本的一些细节更新，如 bug 修复，打补丁等。

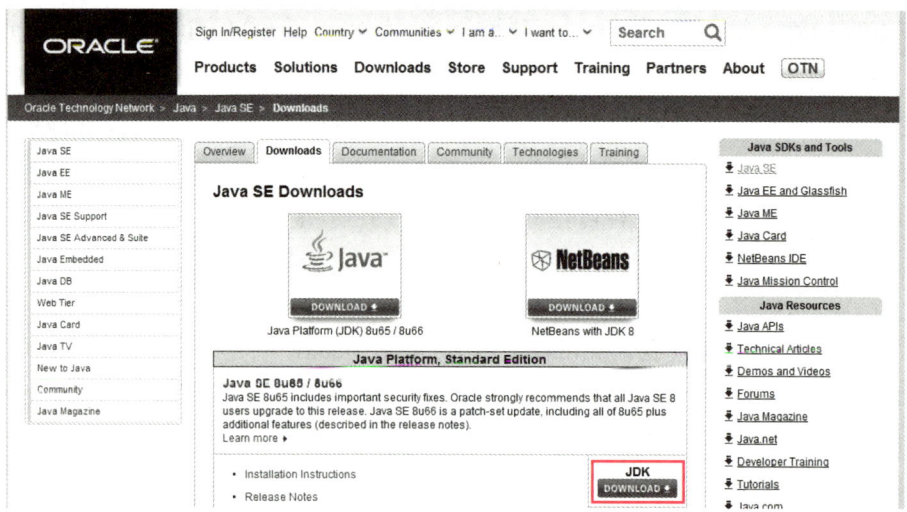

图 2-1　Oracle 官方下载页面

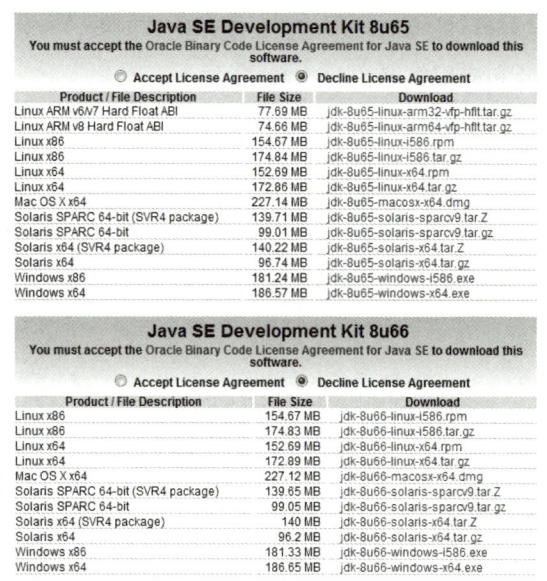

图 2-2　JDK 下载页面

（3）下载完成后双击下载的".exe"文件开始进行安装，将弹出"安装向导"对话框，如图 2-3 所示，在此单击【下一步】按钮。

（4）弹出"安装路径"对话框，如图 2-4 所示，在此选择文件的安装路径。

项目二　Android 开发环境设置

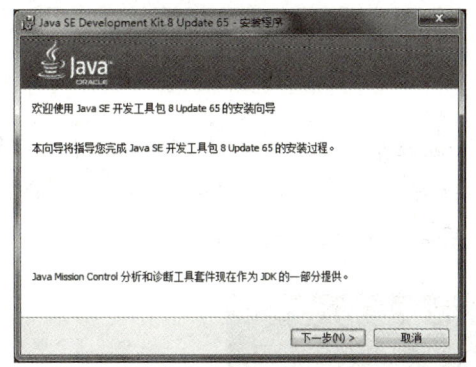 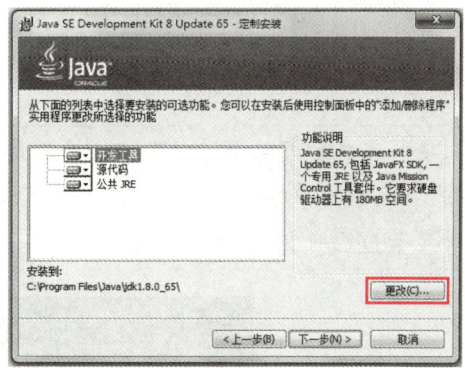

图 2-3　"安装向导"对话框　　　　　图 2-4　"安装路径"对话框

（5）设置安装路径后单击【下一步】按钮，开始在解压缩下载的文件，如图 2-5 所示。完成后弹出"目标文件夹"对话框，在此可以更改安装的位置，如图 2-6 所示。

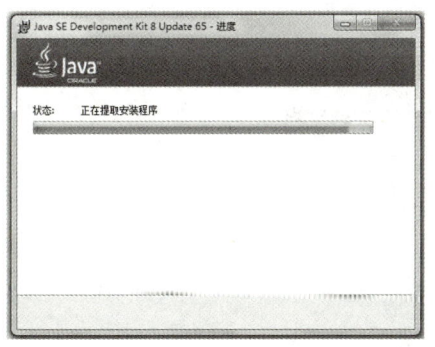 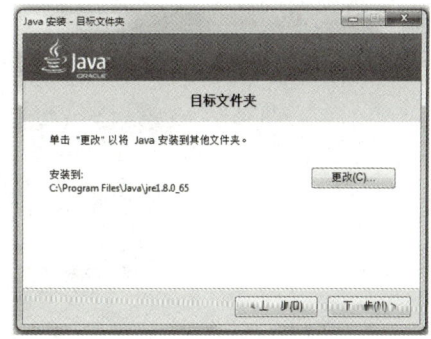

图 2-5　解压缩下载的文件　　　　　图 2-6　"目标文件夹"对话框

（6）单击【下一步】按钮后开始正式安装，如图 2-7 所示。安装完成后弹出"完成"对话框，如图 2-8 所示，单击【关闭】按钮后完成整个安装过程。

 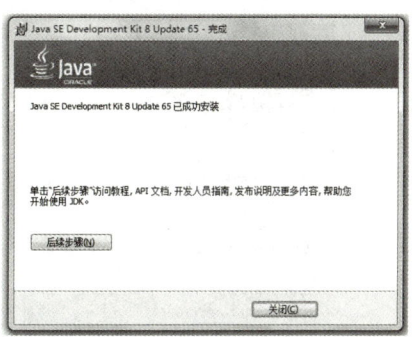

图 2-7　继续安装　　　　　图 2-8　完成安装

> 安装完成后可以通过以下方法检测是否安装成功：依次单击【开始】→【运行】按钮，在运行框中输入"cmd"并按下回车键，在打开的 CMD 窗口中输入 Java –version，如果显示自己所安装的 JDK 版本信息，则说明安装成功，如图 2-9 所示。

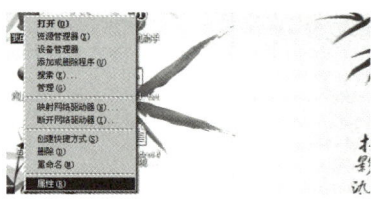

图 2-9　CMD 窗口

JDK 安装完成后还需设定其环境变量，即设置 bin 和 lib 文件夹的路径，这样做的目的是为命令行模式下开发和调试提供便利。具体操作步骤如下：

➢ 新建 JAVA_HOME 变量。

（1）右击我的电脑，在弹出的快捷菜单中选择【属性】选项，如图 2-10 所示。

图 2-10　单击属性

（2）如图 2-11 所示，在打开的"系统属性"对话框中单击"高级"选项卡，然后单击【环境变量】按钮。

（3）在"环境变量"对话框的"系统变量"一栏单击【新建】按钮，如图 2-12 所示。

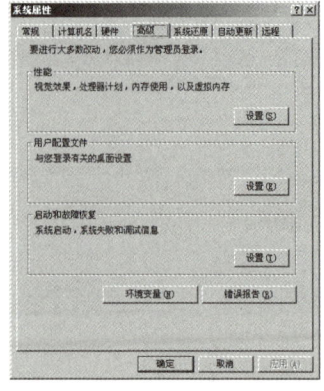

图 2-11　系统属性对话框

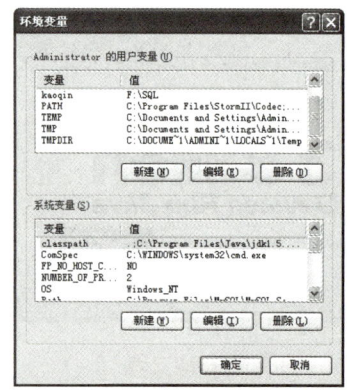

图 2-12　环境变量对话框

（4）在弹出的"新建系统变量"对话框中输入变量名"JAVA_HOME"，变量值为 JDK 的安装位置"C:\Program Files\Java"，如图 2-13 所示，录入完成后单击【确定】按钮。

图 2-13　环境变量对话框

➢ 编辑 Path 变量。

（5）然后参照图 2-12 所示，在系统变量中找到 Path 项，选中后，单击【编辑】按钮，看变量值结尾是否有分号，如果没有，加一个分号隔开，如图 2-14 所示。

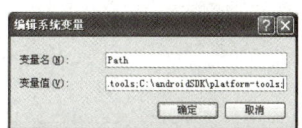

图 2-14　编辑 Path 变量

（6）如图 2-15 所示，将 Java 的 bin 目录除掉 JAVA_HOME 部分添加到 Path 变量的分号后面，单击【确定】保存，如图 2-16 所示。

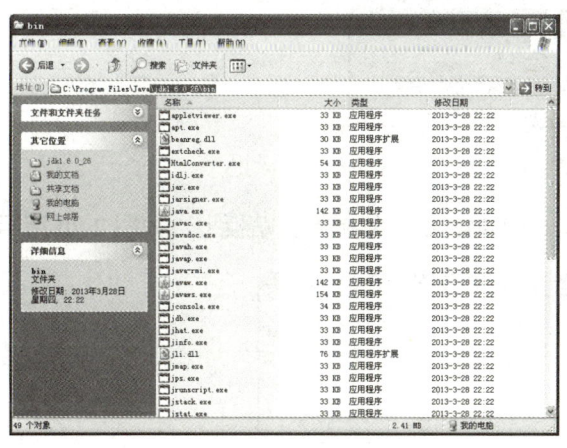

图 2-15　将 Java 的 bin 目录除掉 JAVA_HOME 部分复制到剪贴板上　　图 2-16　添加到分号后面

➢ 新建 CALSSPATH 系统变量

（7）然后参考在系统变量中新建 JAVA_HOME 的操作，新建系统变量"CALSSPATH"。变量名设置为 CLASSPATH（不区分大小写），变量值为一个点代表当前目录，加一个分号隔开，如图 2-17 所示。

图 2-17 新建系统变量 "CALSSPATH"

（8）将 Java 下的 lib 目录除掉 JAVA_HOME 部分拷贝到剪贴板上，如图 2-18 所示。

图 2-18 将 Java 下的 lib 目录除掉 JAVA_HOME 部分拷贝到剪贴板上

（9）添加变量值 "%JAVA_HOME%\jdk1.6.0_26\lib"，然后再加一个分号隔开，如图 2-19 所示。

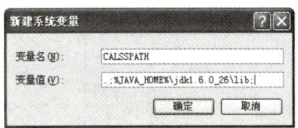

图 2-19 添加变量值 "%JAVA_HOME%\jdk1.6.0_26\lib"

（10）将 jre 下的 lib 目录除掉 JAVA_HOME 部分复制到剪贴板上，如图 2-20 所示。

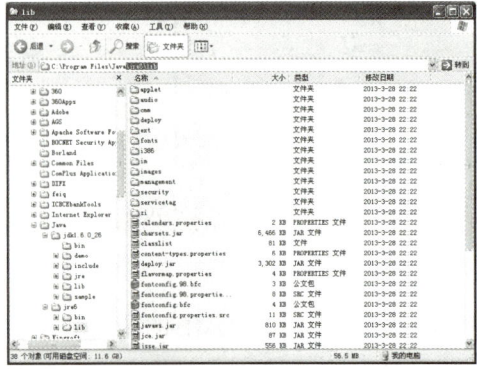

图 2-20 将 jre 下的 lib 目录除掉 JAVA_HOME 部分复制到剪贴板上

（11）添加变量值"%JAVA_HOME%\jre6\lib"，并再加一个分号隔开，如图2-21所示。

图2-21　添加变量值"%JAVA_HOME%\jre6\lib"

（12）添加变量值"%JAVA_HOME%\jdk1.6.0_26\lib\tools.jar"，单击【确定】保存，如图2-22所示。

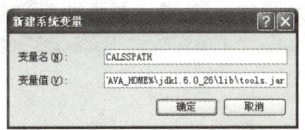

图2-22　添加变量值"%JAVA_HOME%\jdk1.6.0_26\lib\tools.jar"

至此JDK安装完成，验证JDK和环境变量是否安装成功，我们可以在DOS命令框中输入Java和Javac如下两条命令，若显示如图2-23所示，说明我们JDK和环境变量已经安装和配置成功。

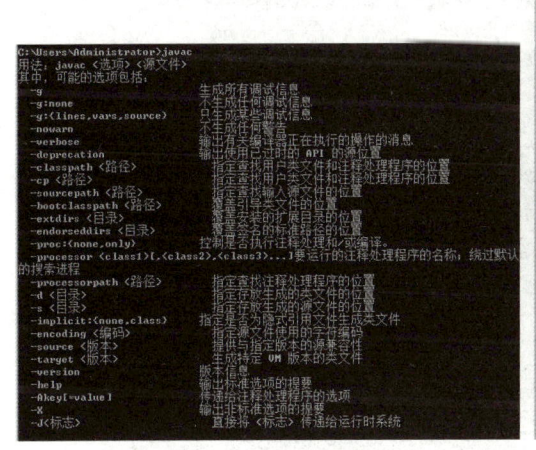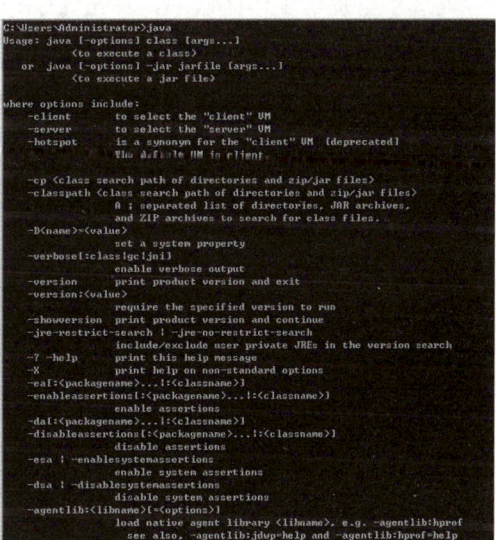

图2-23　运行Javac和Java成功后的效果图

2. 安装Eclipse

安装完成JDK之后，安装Eclipse，步骤如下：

首先，打开Eclipse下载页面，下载网址为http://www.eclipse.org/downloads/，下载最

新版的 Eclipse。

下载完成之后，解压 Eclipse 压缩包，然后进入解压目录内就会看到一个名为 eclipse.exe 的可执行文件，双击 eclipse.exe 直接启动 Eclipse。如果读者是第一次启动 Eclipse，将会看到一个选择工作区（Workspace）的提示，如图 2-24 所示。

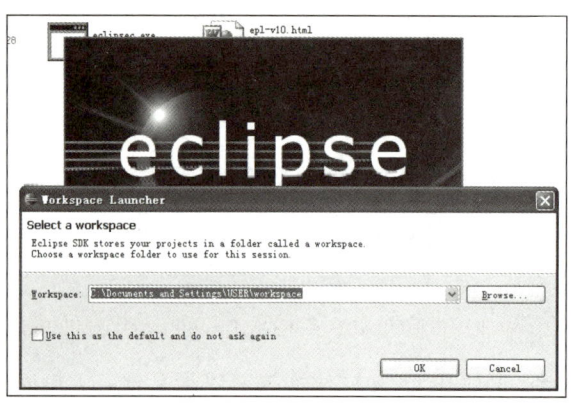

图 2-24　启动 Eclipse

勾选不再提示工作区选项，单击【OK】按钮打开 Eclipse，Eclipse 打开成功进入欢迎界面，如图 2-25 所示。

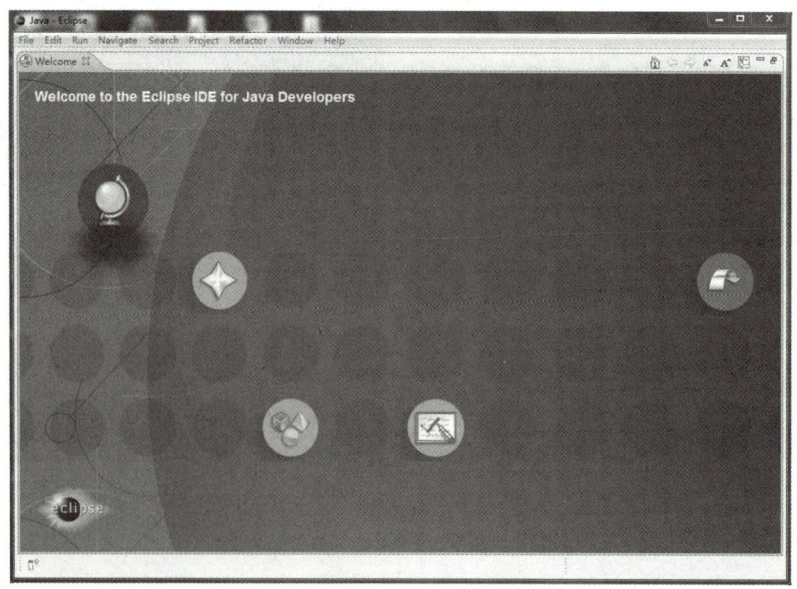

图 2-25　Eclipse 欢迎界面

3. 安装 SDK

Android SDK 的官方网站为 http://developer.android.com/sdk/index.html，可以从该网站下载最新版的 Android SDK，如图 2-26 所示。

Platform	Package	Size	MD5 Checksum
Windows	android-sdk_r12-windows.zip	36486190 bytes	8d6c104a34cd2577c5506c55d981aebf
	installer_r12-windows.exe (Recommended)	36531492 bytes	367f0ed4ecd70aefc290d1f7dcb578ab
Mac OS X (intel)	android-sdk_r12-mac_x86.zip	30231118 bytes	341544e4572b4b1afab123ab817086e7
Linux (i386)	android-sdk_r12-linux_x86.tgz	30034243 bytes	f8485275a8dee3d1929936ed538ee99a

图 2-26　Android SDK 的官方网站

这里我们选择下载适合 Windows 平台开发的 Android SDK 程序包，包名为 android-sdk_r16-windows.zip，下载完成之后解压缩到我们的工作目录内，如：F:\google_android\Java_windows\。在 SDK 目录内我们会发现，Android SDK 不再捆绑 platform 和 add-on，因此这两部分需要手动下载。执行 SDK 里附带的"SDK Manager.exe"程序，如图 2-27 所示。

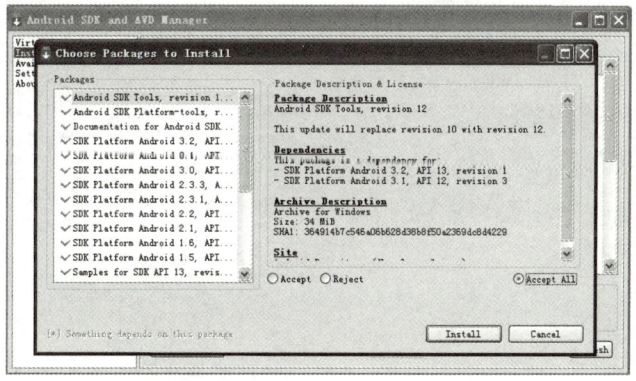

图 2-27　下载工具包

这里我们选择【Accept All】下载所有的程序，然后单击【Install】按钮即可开始下载安装，如图 2-28 所示。整个下载过程需要很长一段时间，读者需要耐心等待。

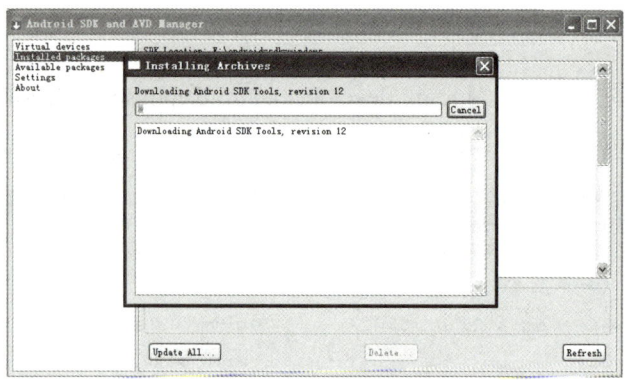

图 2-28　SDK 工具包安装过程

安装完成之后，退出该页面。接下来我们要做的就是对 SDK 进行配置。需要将 Android SDK 目录中的 tools 和 platform-tools 文件夹路径添加到 PC 的环境变量 PATH 中，具体配置过程在此就不再讲述，请读者参考 JDK 环境配置的过程。

重新启动 Eclipse 之后，需要在 Eclipse 的 Preferences 中添加 Android SDK 的路径。单击【Windows】→【Preferences】选项，如图 2-29 所示。

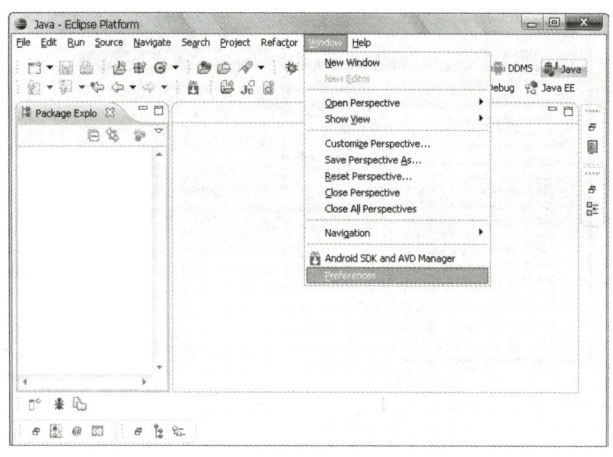

图 2-29　添加 SDK 路径

在当前对话框左侧选择"Android"项，然后单击右侧的【Browse】按钮选择 Android SDK 的路径，如图 2-30 所示。

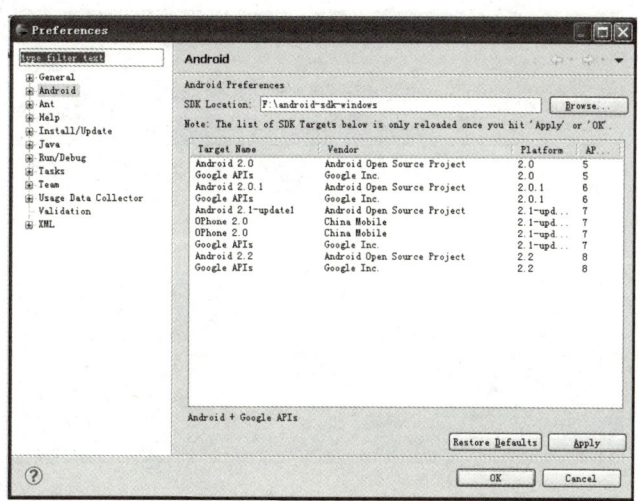

图 2-30　添加 SDK 路径

添加成功之后单击【Apply】按钮载入 SDK 包，载入完成后，单击【OK】按钮退出即可。

4. 安装 ADT

成功安装 Eclipse 之后，还需要安装 ADT 开发工具，因为 ADT 是替 Eclipse 打造一个 Android 专属的开发环境，包括创建 Android 开发实例、执行和除错，全部添加到 Eclipse 整合开发环境中，这样就不需要再使用 adb.exe 除错指令进行除错了。

（1）打开 Eclipse 后，单击菜单栏中的【Help】→【Install New Software...】选项，如图 2-31 所示。

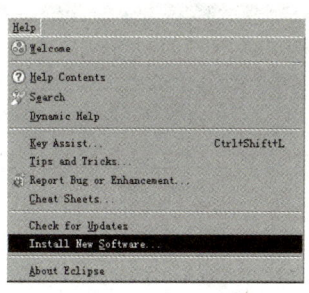

图 2-31　添加插件

（2）在弹出的"Install"对话框中单击【Add】按钮，如图 2-32 所示。

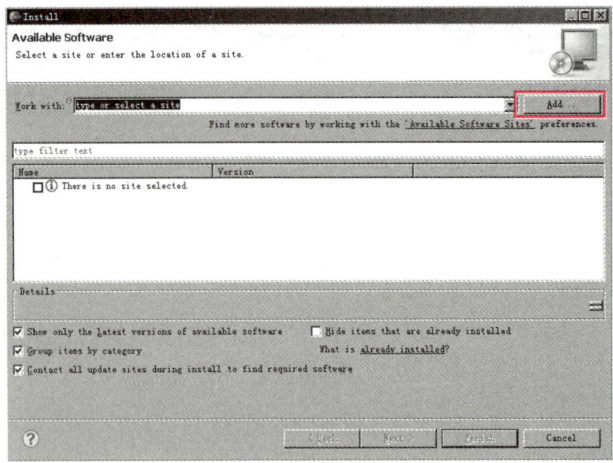

图 2-32 添加插件

（3）在弹出的"Add Site"对话框中分别输入名字和地址，名字可以自己命名，但是 Location 中需输入插件的网络地址 http://dl-ssl.google.com/Android/eclipse，如图 2-33 所示。

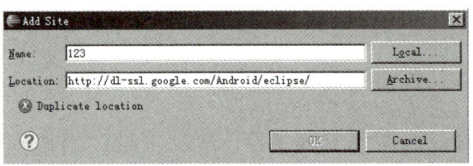

图 2-33 设置地址

（4）单击【OK】按钮，此时在"Install"对话框中将会显示系统中可用的插件，如图 2-34 所示。

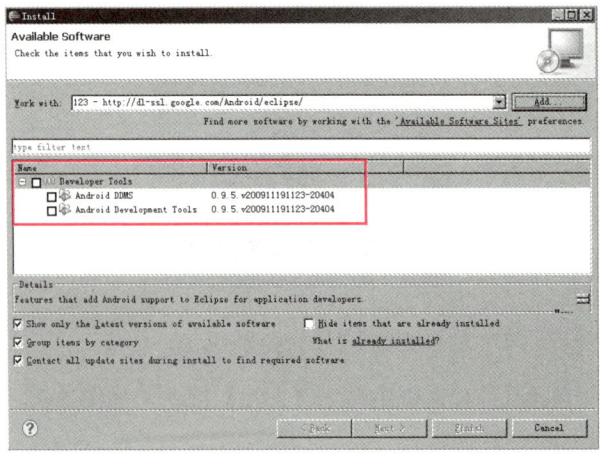

图 2-34 插件列表

项目二　Android 开发环境设置

在笔者完成此书时，Google 在我国还处于被限制访问状态，访问和下载 Android 官方网站的资料（Android 是 Google 公司产品）变得非常困难。如果不能显示系统中可用的插件或者显示 pending，建议读者使用代理服务器或寻找镜像文件进行下载。

（5）选中"Android DDMS"和"Android Development Tools"，然后单击【Next】按钮来到如图 2-35 所示的安装界面。

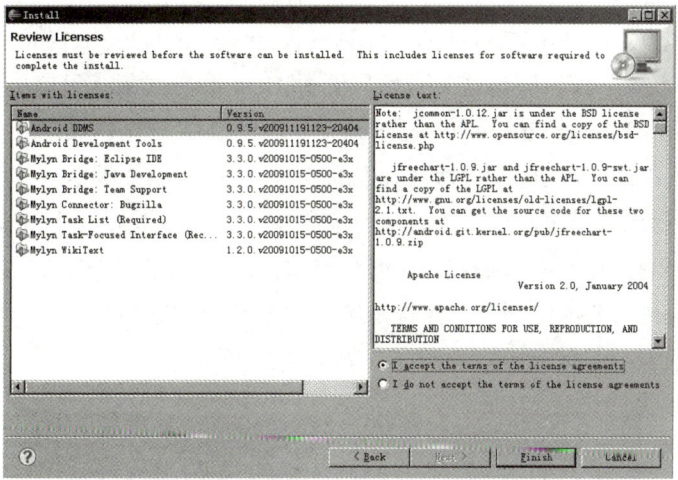

图 2-35　插件安装界面

（6）选择"I accept"选项后单击【Finish】按钮即开始进行安装，如图 2-36 所示。

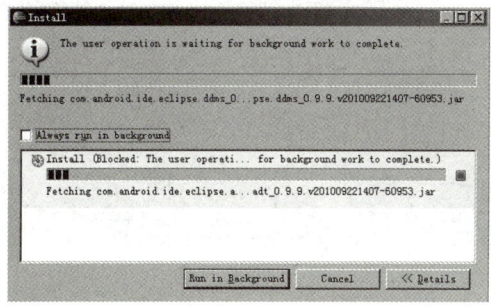

图 2-36　开始安装

在上个步骤可中，可能会发生计算插件占用资源情况，过程有点慢。完成后会提示重启 Eclipse 来加载插件，等电脑重启后就可以使用了。不同版本的 Eclipse 安装插件的方法和步骤是不同的，但是大同小异，读者可根据操作提示自行安装。

5. Android Studio 的下载、安装和使用

作为一个 Android 开发者，可以在 Google Android 的一些官方网站下载安装 Android Studio，有两个地方：Android Developer 官网和 Android Tools Project Site。Android 开发者官网的网站，可直接下载，但是这个网站只更新 Beta 和正式版，目前更新到 Beta 0.8.14 版本。Android Tools Project Site 是 Android 开发工具的网站，给出的链接是 Studio 的 canary 渠道，列出了 Studio 各种实时预览版等，目前最新的是 1.0 RC 版本。

至于安装过程比较简单，单击直接运行安装即可。这里以 Mac 系统的 1.0 RC 版本为例，来创建第一个 HelloWorld 项目。其他平台的操作基本上差不多，在这之前假设已经配置好了 JDK 和 Android SDK 环境，并且是第一次安装 Studio。

（1）首先显示的是运行时欢迎画面，如图 2-37 所示。

图 2-37　欢迎画面

（2）之后进入到设置向导页面，如图 2-38 所示。

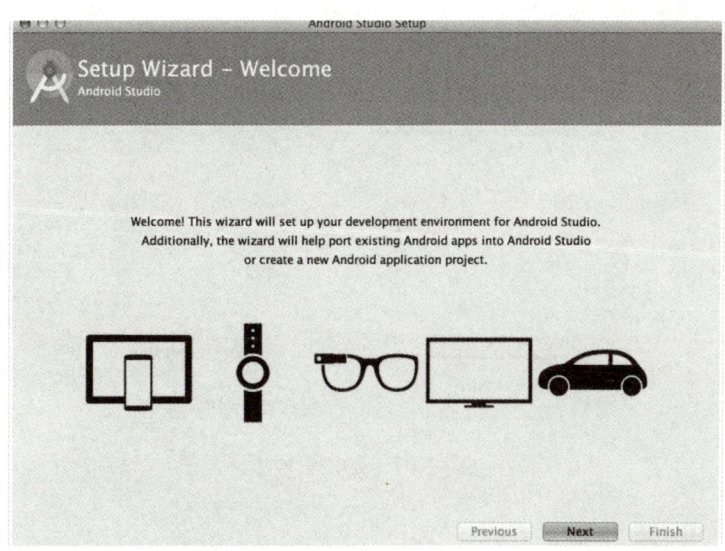

图 2-38 设置向导

(3) 单击【Next】按钮进入安装类型选择向导页,如图 2-39 所示。

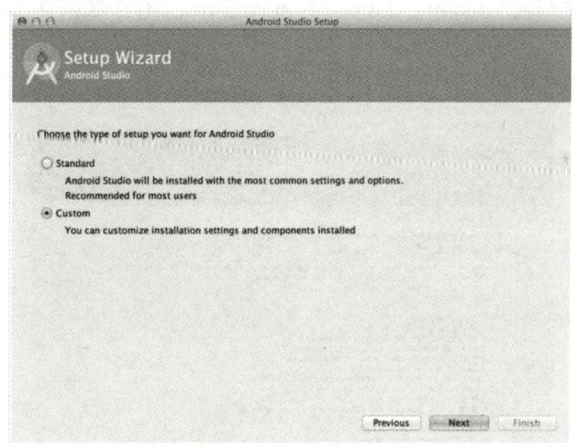

图 2-39 安装向导

这里有两个选项"Standard"和"Custom",即标准和自定义,如果本机的 Android SDK 没有配置过,那么建议直接选择"Standard"选项,单击【Finish】按钮。这里因为本地已经下载 SDK 并配置好了环境变量,因此选择"Custom"选项,然后单击【下一步】按钮。

(4) 下面选择本地 SDK 的位置,如图 2-40 所示,可以看到有个 2.25 GB 的 SDK 要下载,那是因为 Studio 1.0 默认要下载 5.0 的 SDK 以及一些 Tools 等,单击【Finish】按钮。

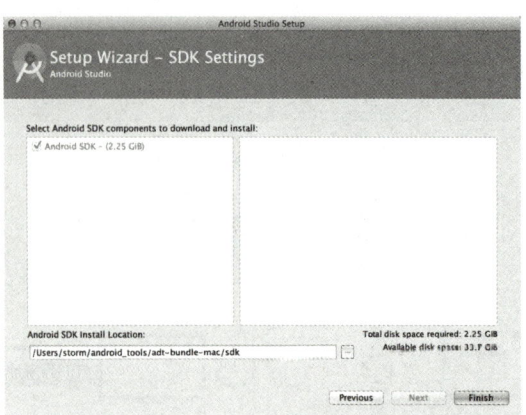

图 2-40　SDK 设置

　　选择并下载 2.25 G 的组件不是必须进行。如果想跳过这步的可以在 bin 目录的 idea.properties 增加一行：disable.android.first.run=true 就可以。

（5）接下来下载依赖的组件，如图 2-41 所示。

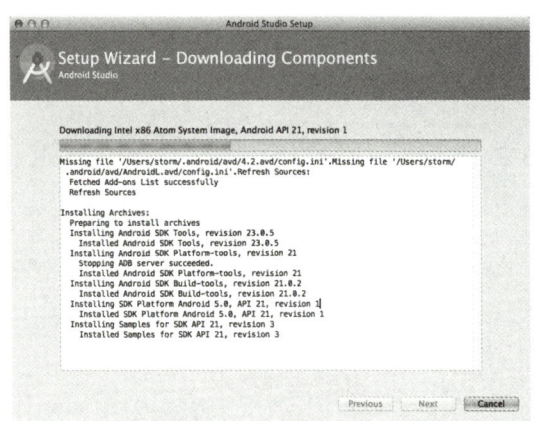

图 2-41　下载依赖组件

　　这个过程需要等待，而且依赖网速，时间可能会有点久。

（6）下载完成后如图 2-42 所示，单击【Finish】按钮即可。

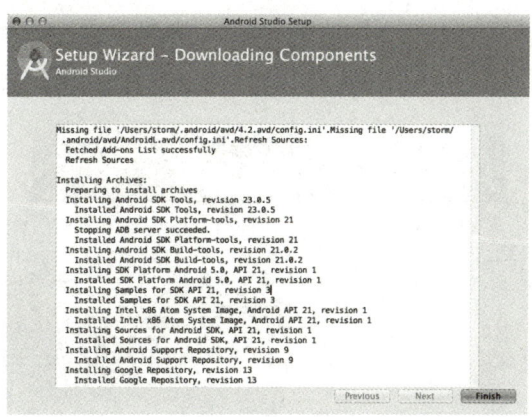

图 2-42　下载组件完成

（7）接下来的页面如图 2-43 所示。在这个页面我们可以新建项目，也可以导入本地或者 GitHub 上的项目等，左边可以查看最近打开的项目，这里直接新建项目。

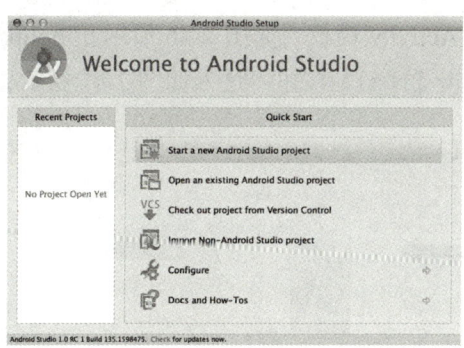

图 2-43　新建项目

（8）进入图 2-44 所示界面，填上项目名称、包名以及项目路径，单击【Next】按钮。

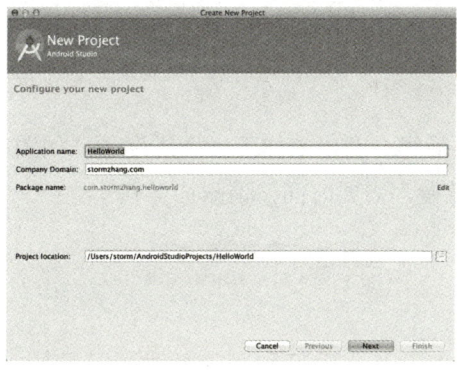

图 2-44　新项目参数

（9）这个页面支持适配 TV、Wear、Glass 等，只选择第一项即可，选好最低 SDK 版本，如图 2-45 所示，然后单击【Next】按钮。

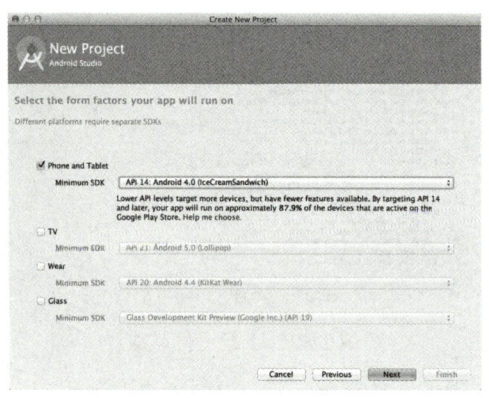

图 2-45　设备及 SDK 参数

（10）下面选择一个 Activity 模板，如图 2-46 所示和 Eclipse 很像，我们直接选择一个 Blank Activity，然后单击【Next】按钮，进入如图 2-47 所示界面填写 Activity 信息。

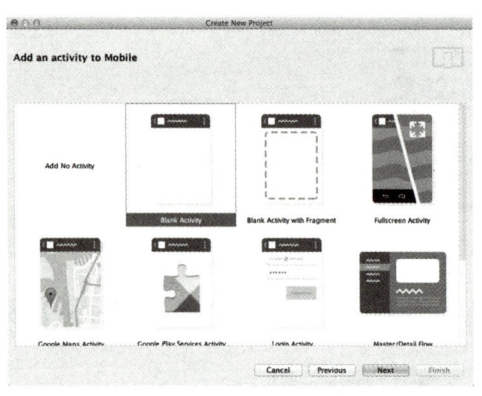 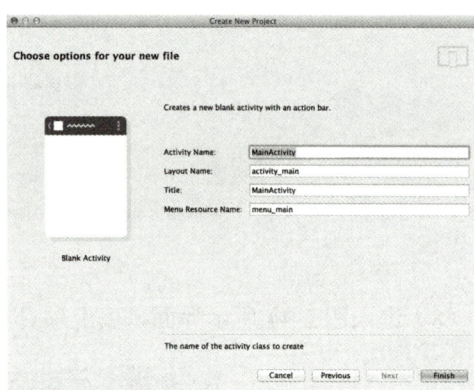

图 2-46　选择 Activity 模板　　　　　　　　图 2-47　文件设置

（11）单击【Finish】后等一会出来如下一个进度条，如图 2-48 所示，这里需要下载 Gradle，第一次下载会有点慢，需要时间，请耐心等待下。

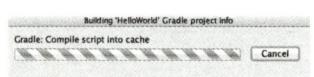

图 2-48　下载 Gradle

（12）下载成功后就会看到如下完整的项目界面，如图 2-49 所示。

项目二　Android 开发环境设置

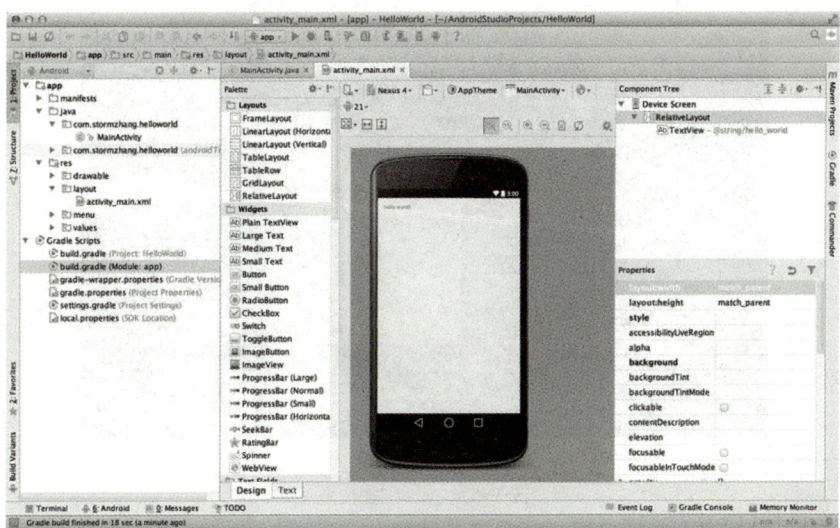

图 2-49　默认项目界面

一个简单的 Android Studio 项目完成了，图片中也可以看到默认是一个白色主题。

（13）如果要使用 Studio 默认自带的一款黑色主题，只需要简单修改下就可以。可以在【Preference】→【Appearance】下更改主题到 Darcula，如图 2-50 所示。

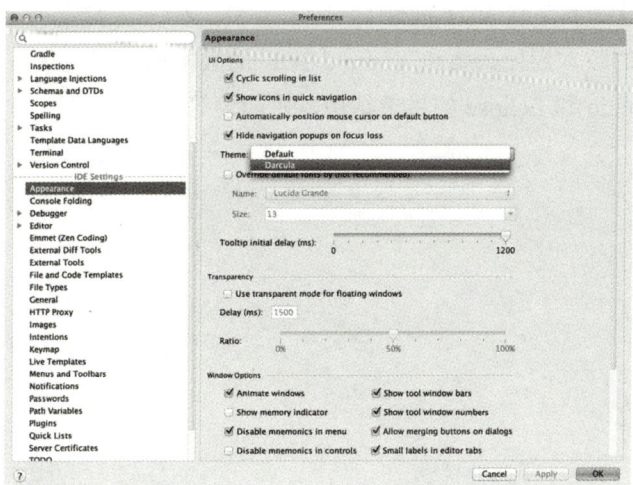

图 2-50　新主题效果

（14）更改主题后如图 2-51 所示。

31

图 2-51　新主题效果

至此 Android Studio 的安装介绍完毕。

四、项目思考与扩展

1. Eclipse 为什么要安装 ADT 插件？
2. ADT 插件的版本为什么要和 Android SDK 的版本相匹配？
3. Android Studio 和 Eclipse 的目录结构、快捷键等有哪些不同？

项目三 Android 模拟器与开发终端

【本章导读】

在开发 Android 项目时,我们需要对所开发的应用程序进行安装和调试,开发人员可采用 AVD(Android Virtual Device)即 Android 模拟器,在个人电脑上进行 APP 程序的调试,也可以使用真实的 Android 设备进行开发和调试。为了降低开发 Android 应用的成本,Android SDK 中提供了一个模拟器,在计算机中开发的应用程序都可以在其中进行测试。本章的学习项目就是掌握如何创建 Android 模拟器和如何连接 Android 真实设备并使用其进行开发调试。

一、项目要求

1. 掌握模拟器的使用。
2. 掌握应用程序在模拟器上的安装和卸载。
3. 掌握应用程序在移动终端的安装和卸载。

二、项目相关知识

1. Android 模拟器简介

所谓模拟器,就是指在电脑上模拟安卓系统,通过模拟器来调试并运行开发的 Android 程序。开发人员不需要一个真实的 Android 手机,只通过电脑即可模拟运行一个手机,开发出应用在手机上面的程序。

AVD 全称为 Android Virtual Device（Android 虚拟设备），每个 AVD 模拟了一套虚拟设备来运行 Android 平台，这个平台至少要有自己的内核，系统图像和数据分区，还可以有自己的 SD 卡和用户数据以及外观显示等。程序开发人员可以根据实际的开发需要，建立多个 AVD。

一个 AVD 由以下几个部分组成：

（1）硬件配置：定义虚拟设备的硬件特征，如程序开发人员可以定义设备是否含有摄像头、定义内存和外设的大小等。

（2）系统的镜像：开发人员可以定义虚拟设备运行的 Android 平台的版本。

（3）其他选项：开发人员也可以定义所使用模拟器的皮肤等。

（4）计算机的存储专区：用于存储当前设备的用户数据。

以 Android 4.0 系统的 AVD 为例，Android 模拟器的主要功能如下：

（1）应用程序按钮，单击该按钮会显示系统安装的应用程序。

（2）设备状态，包括时间、信号强度、电量等。

（3）项目切换键，单击后显示最近运行的程序。

（4）Home 键，用于返回桌面。

（5）后退键，用于返回前一个应用。

（6）显示当前所安装的程序。

启动 Android 模拟器，有以下 3 种常见的方式：

（1）使用 AVD 管理工具。

（2）使用 Eclipse 运行 Android 程序。

（3）使用 emulator 命令。

2. AVD 和真实设备的区别

虽然目前 Android 模拟器有许多应用功能，可以模拟真实移动设备上的许多硬件和软件特性，但是与真实的 Android 真机相比较，还是存在着一些局限性，下面一些功能是 AVD 目前无法模拟的：

（1）不具备接听和呼叫真实设备的性能。

（2）不支持 USB 连接。

（3）不支持相机的功能。

（4）不支持语音的输入和输出。

（5）不支持扩展耳机的功能。

（6）不支持确定实际电量的功能。

（7）不能模拟 SD 卡的实际插入语插出。

（8）不具备蓝牙和 WiFi 与重力感应等功能。

三、项目实施过程

开发 Android 应用程序，需要手动建立虚拟设备或使用真实设备进行开发。下面我们分别来了解这两种方式的操作。

1. 创建 Android 虚拟设备

我们在开发时，按照如下的步骤进行 Android 虚拟设备的创建。

（1）单击 window 项，在 window 项的子项中选择【AVD Manager】项，如图 3-1 所示。

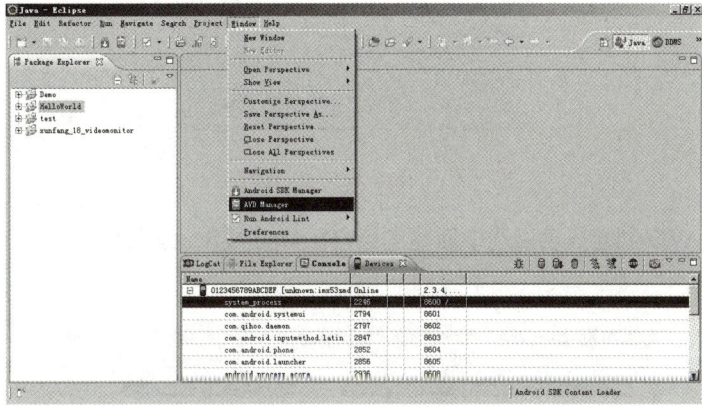

图 3-1　单击 AVD Manager 项

（2）在 AVD Manager 对话框中，单击右侧的【New...】按钮，如图 3-2 所示。

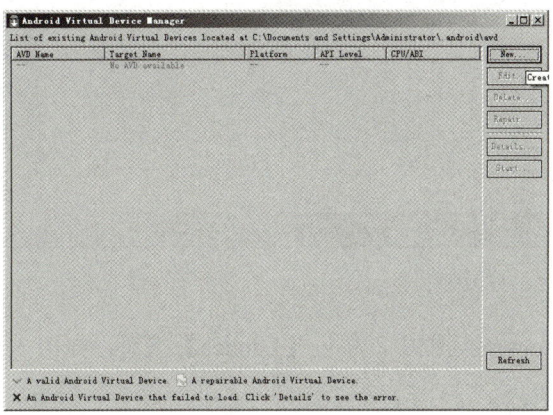

图 3-2　AVD Manager 对话框

(3) 在 Create New AVD 对话框中输入 Name, 选择 Target, 输入 SD Card Size, 然后单击【Create AVD】按钮, 如图 3-3 所示。

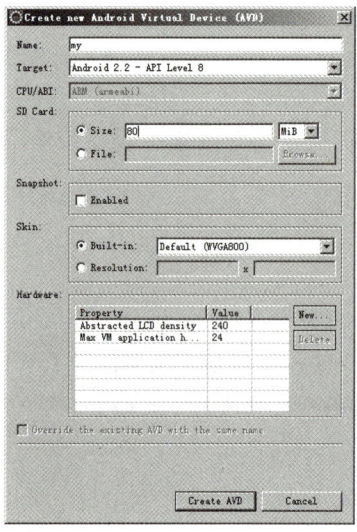

图 3-3　Create New AVD 对话框

(4) 至此 AVD 创建成功。选中新建的 AVD, 单击【Start】按钮启动虚拟设备, 如图 3-4 所示。

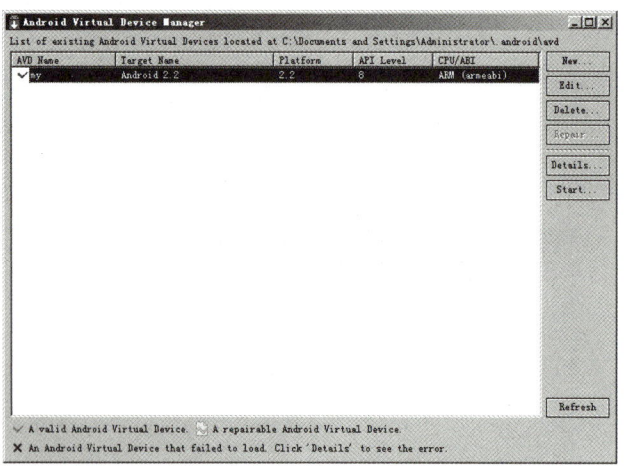

图 3-4　单击【Start】按钮

(5) Launch Options 对话框中, 单击【Launch】按钮, 如图 3-5 所示。

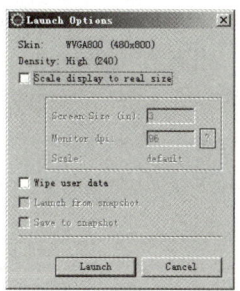

图 3-5　单击【Launch】按钮

（6）AVD 启动后的虚拟设备界面如图 3-6 所示，这样我们就将 Android 虚拟设备建立完成了。

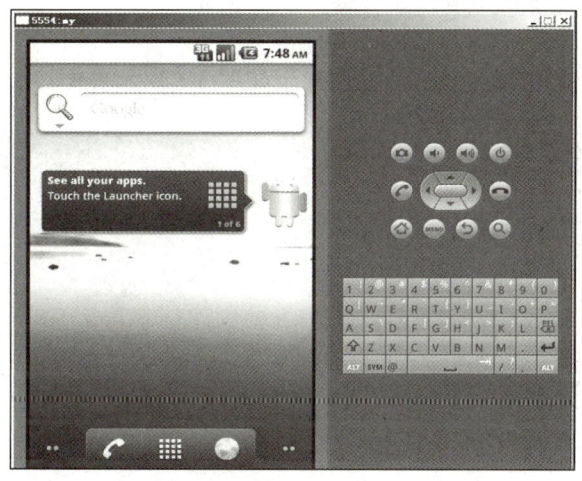

图 3-6　Android 模拟设备

2. 连接真实 Android 设备

在 Android 程序开发中，我们除了可以选择模拟器开发外，也可以使用真实设备进行开发和调试，具体方法如下：

（1）首先将手机设置为调试模式，具体操作如下：

【设置】→【应用程序】→【开发】→【USB 调试】，打上√即可。

（2）用数据线连接至电脑，在电脑上安装豌豆荚，此时豌豆荚会帮你安装驱动，安装好后豌豆荚就可以连接上手机了。

（3）开始在真机上调试，在 eclipse 中选择【Run】→【Run Configurations】项，在左边选择好所要调试的工程，然后将右边切换至 Target 标签下，如图 3-7 所示。

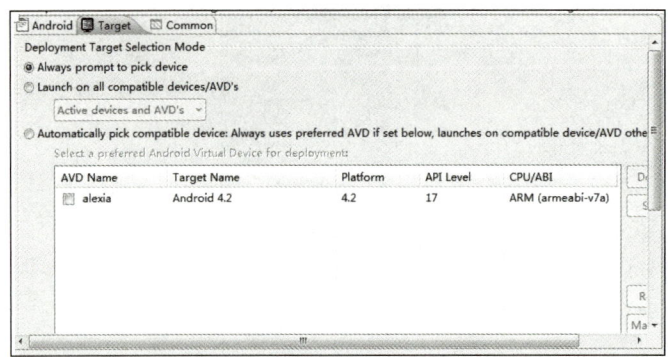

图 3-7　真机调试界面

这有三个选项，如果你想连接至真机调试，可选第一个或第二个，这里我直接选择第一个。

（4）单击【Run】按钮，等待几秒钟出现如图 3-8 所示界面。

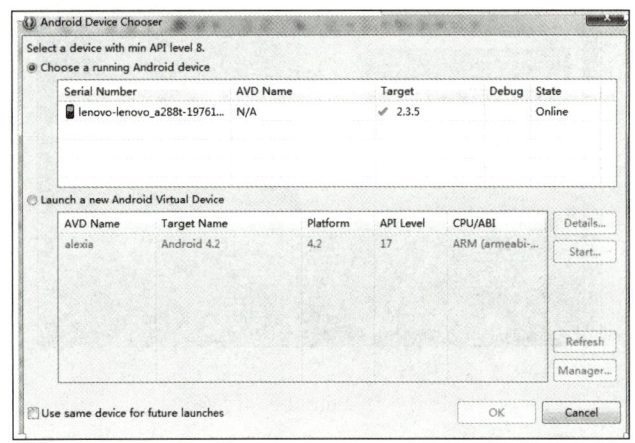

图 3-8　真机调试界面运行界面

在这里就可以看到我们的真机装置了，选择上面的真机，即可运行程序。

四、项目思考与扩展

分别使用 Android 虚拟设备和真实设备练习 APP 应用程序的安装与卸载。

项目四　Android 工程创建与资源使用

【本章导读】

在前面的项目中，我们已经学习了 Android 开发环境的搭建与 Android 虚拟设备的创建，在本章中，我们将创建第一个 Android 程序，作为程序员，学习一门新语言的第一步就是输出 "Hello World"。

一、项目要求

1. 掌握创建 Android 工程的方法。
2. 掌握 Android 终端联机调试方法。
3. 了解 Layout、AndroidManifest.xml 等文件的用途。

二、项目相关知识

1. Android SDK 软件开发包

SDK（Software Development Kit）软件开发工具包是软件开发工程师用于为特定的软件包、软件框架、硬件平台、操作系统等建立应用软件的开发工具的集合。由此，Android SDK 指的是 Android 专属的软件开发工具包。

在 Android SDK 中，存在着许多目录文件夹，如图 4-1 所示。

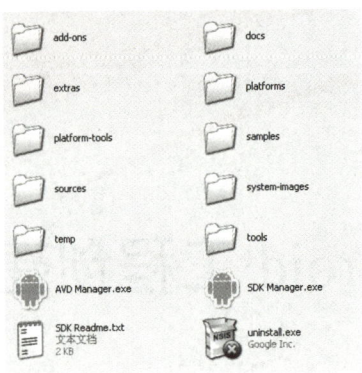

图 4-1　Android SDK

这些文件夹的作用如下：
- add-ons 目录用来存放 Google 提供的地图开发包，支持基于 Google Map 的地图开发。
- docs 目录中存放的是 Android SDK 的帮助文档，通过目录下的 offline.html 文件启动。
- extras 目录下保存了 Android 手机的 USB 驱动程序。
- platforms 目录用来存放 SDK 和 AVD 管理器下载的各种版本的 SDK。
- platforms-tool 目录中保存了与平台调试相关的工具。
- Samples 目录是示例代码和程序的存放目录。
- Temp 是临时存放文件的目录。
- tools 目录保存了通用的 Android 开发调试工具和 Android 手机模拟器。
- SDK Manager.exe 和 AVD Manager.exe 分别是 SDK 和 AVD 的管理器。
- SDK Readme.txt 是 Android SDK 的说明文档。

2. Android 程序结构

当我们新建一个 Android 工程后，ADT 会以工程名作为根目录，将所有自动生成和非自动生成的文件都保存在这个根目录下。根目录下包含 5 个子目录 src、gen、assets、bin 和 res，1 个库文件 android.jar，以及 3 个工程文件 AndroidManifest.xml、project.properties 和 proguard.cfg，如图 4-2 所示。

下面我们介绍一下 Android 应用程序组成文件的具体信息，这对于我们后边的学习有重要的指导作用。

项目四　Android 工程创建与资源使用

```
HelloWorld
├── src
│   └── com.xunfang.helloworld
│       └── HelloWorldActivity.java
├── gen [Generated Java Files]
│   └── com.xunfang.helloworld
│       ├── BuildConfig.java
│       └── R.java
├── Android 2.2
│   └── android.jar - D:\android-sdk-windows\platforms\android-8
├── Android Dependencies
├── assets
├── bin
├── res
│   ├── drawable-hdpi
│   ├── drawable-ldpi
│   ├── drawable-mdpi
│   ├── drawable-xhdpi
│   ├── layout
│   └── values
├── AndroidManifest.xml
├── proguard-project.txt
└── project.properties
```

图 4-2　Android 程序结构

- src 目录是源代码目录，所有允许用户修改的 Java 文件和用户自己添加的 Java 文件都保存在这个目录中。
- gen 目录用来保存 ADT 自动生成的 Java 文件。
- Android 2.2 这个目录是用来存放 Android 自身的所有 jar 包文件。建立不同版本的可能会有不同的依赖。Android.jar 文件是 Android 程序所引用的函数库文件，Android 系统所支持的 API 都包含在这个文件中。
- assets 目录用来存放原始格式的文件。
- bin 目录保存了编译过程中产生的文件，以及最终生成的 apk 文件。
- res 目录是资源目录，Android 程序所有的图像、颜色、风格、主题、界面布局和字符串等资源都保存在其下的几个子目录中。其中：drawable-hdpi、drawable-mdpi 和 drawable-ldpi 目录用来保存同一个程序中针对不同屏幕尺寸需要的不同大小的图像文件，layout 目录用来保存与用户界面相关的布局文件，values 目录保存颜色、风格、主题和字符串等资源。
- AndroidManifest.xml 该文件是每个 Android 项目必须的系统控制文件。它向 Android 操作系统描述了本程序所包括的组件、所实现的功能、能处理的数据、要请求的资源等等。
- proguard-project.txt 该文件是混淆代码的脚本配置文件。
- project.properties 该文件存储当前应用所使用的 android 配置信息。

3. DDMS

DDMS（Dalvik Debug Monitor Service）是 Android 系统中内置的调试工具，如图 4-3 所示。它可以用来监视 Android 系统中的进程、堆栈信息，查看 logcat 日志，实现端口转发服务和屏幕截图功能,模拟电话呼叫和 SMS 短信,以及浏览 Android 模拟器文件系统等。

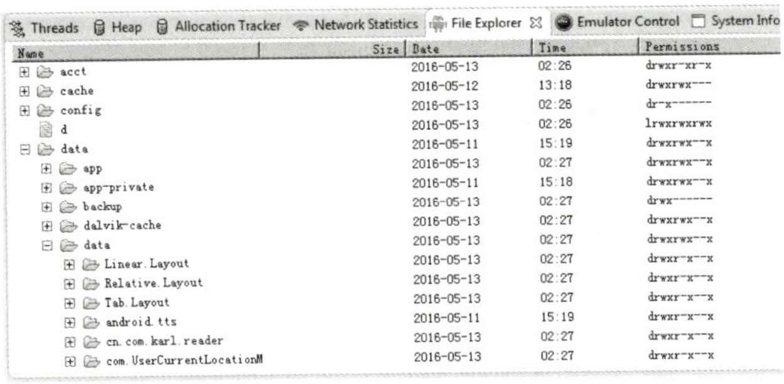

图 4-3　DDMS

三、项目实施过程

参照项目二和项目三的操作步骤,运行 Eclipse,启动 Android 开发终端,并启动 Android 模拟器进行第一个 Android 项目 HelloWorld 的开发。项目实施过程如下:

1. 建立 Android 项目

(1)新建一个 Android 工程,在菜单栏中执行【File】→【New】→【Project】命令,如图 4-4 所示。

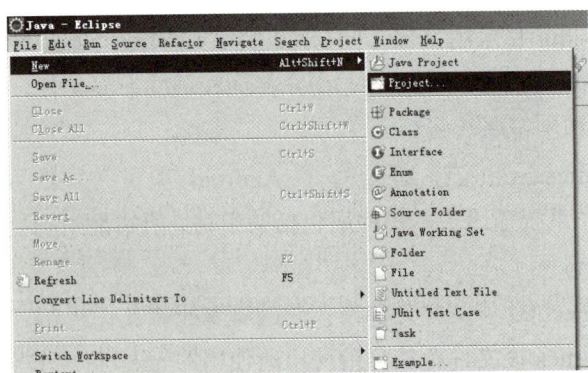

图 4-4　单击【Project】项

(2)在 New Project 对话框中,选中 Android 目录下的 Android Project,然后单击【Next】按钮,如图 4-5 所示。

项目四　Android 工程创建与资源使用

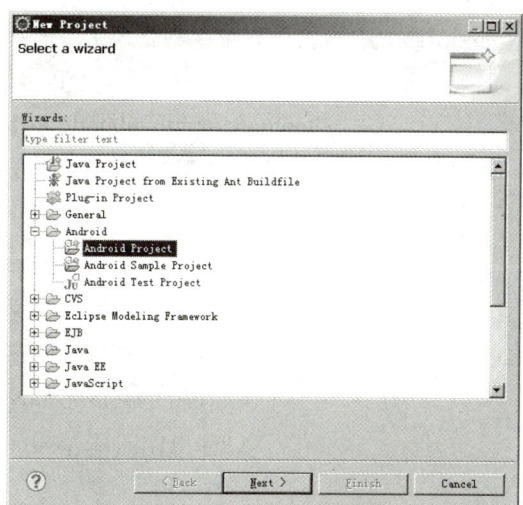

图 4-5　New Project 对话框

（3）在 New Android Project 对话框中，输入工程名"HelloWorld"，然后单击【Next】按钮，如图 4-6 所示。

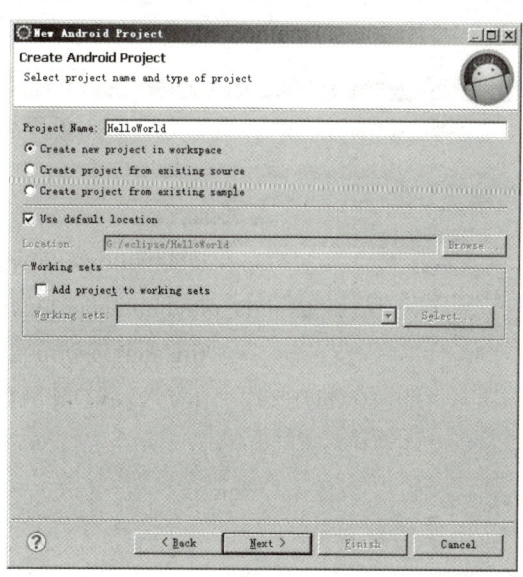

图 4-6　New Android Project 对话框

（4）在 Android 版本选择对话框中，选择 Android 版本为 2.2，如图 4-7 所示。

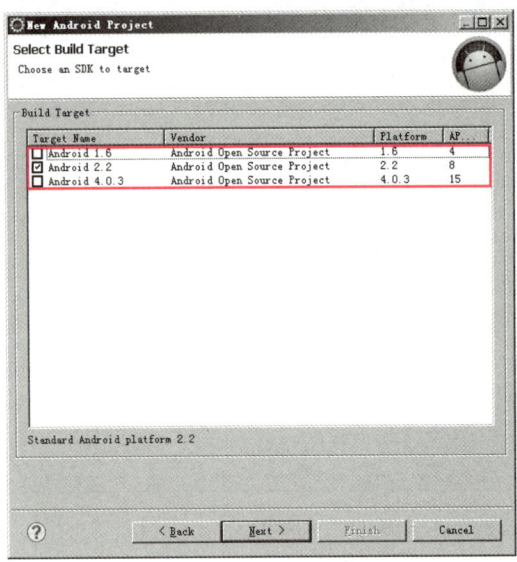

图 4-7　选择 Android 版本

（5）在应用信息配置界面输入包名"com.my.first"，单击【finish】按钮，如图 4-8 所示。

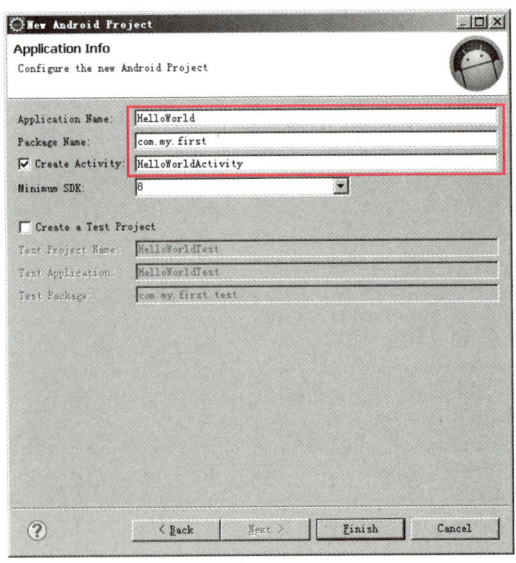

图 4-8　应用信息配置

（6）此时便新建了一个名为"HelloWorld"的工程，如图 4-9 所示。

项目四　Android 工程创建与资源使用

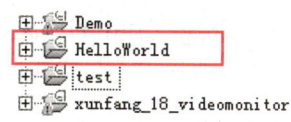

图 4-9　新建的工程

如果 Eclipse 右侧的 Package Explorer 窗口未打开，则单击菜单栏中的【Window】→【Show View】→【Other】项，如图 4-10 所示。接着在 Show View 对话框中，打开 Java 目录，如图 4-11 所示，找到此目录下的 Package Explorer 项，双击打开即可。

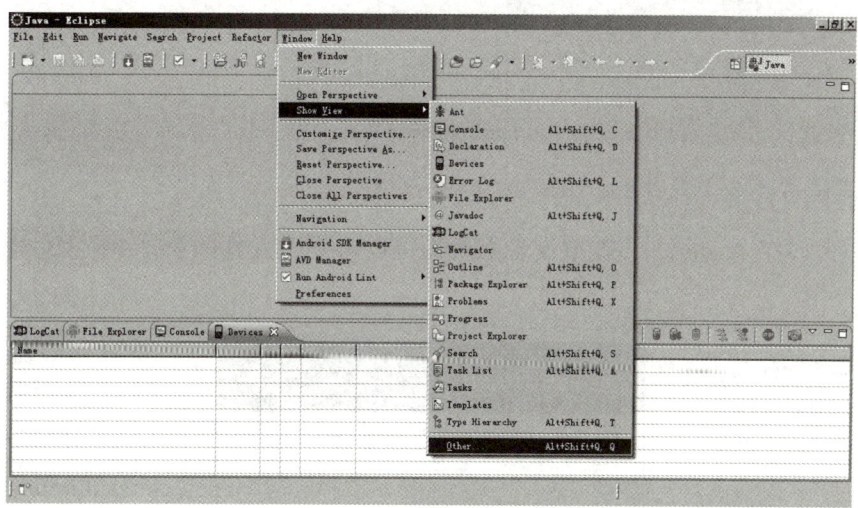

图 4-10　单击 Other 项

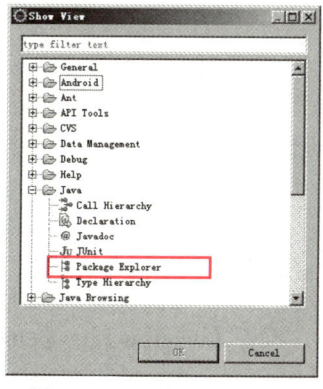

图 4-11　Show View 对话框

（7）右击刚刚新建的 HelloWorld 工程，在弹出的快捷菜单中单击【Run As】→【Android Application】项，如图 4-12 所示。

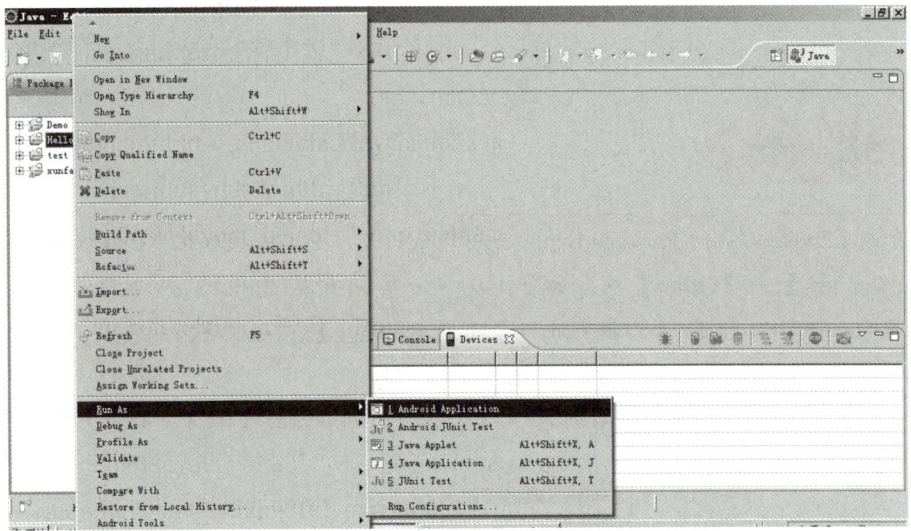

图 4-12　单击 Android Application 项

（8）此时在 Andorid 的虚拟设备上我们可以看到建立好的第一个项目，如图 4-13 所示。

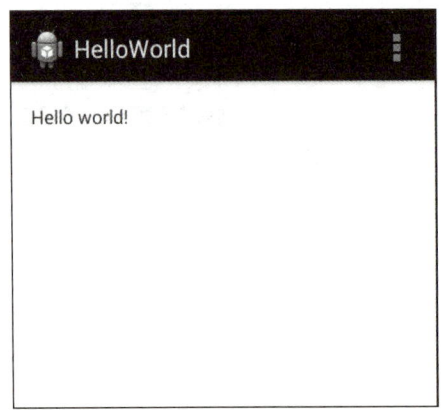

图 4-13　项目在 Andorid 虚拟机上运行的效果

2．DDMS 的使用

（1）在 Eclipse 的菜单栏中，执行【window】→【Open Perspective】→【DDMS】命令，如图 4-14 所示。

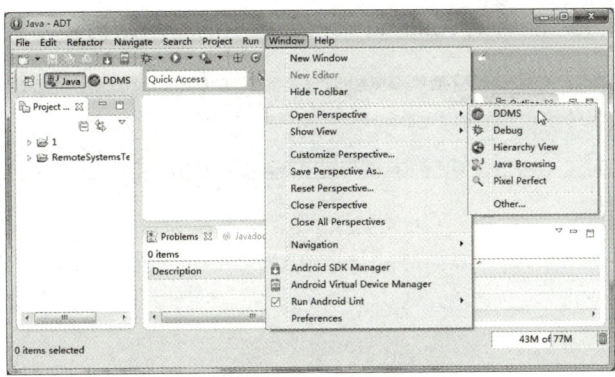

图 4-14　打开 DDMS

此时将会看到如图 4-15 所示的界面。

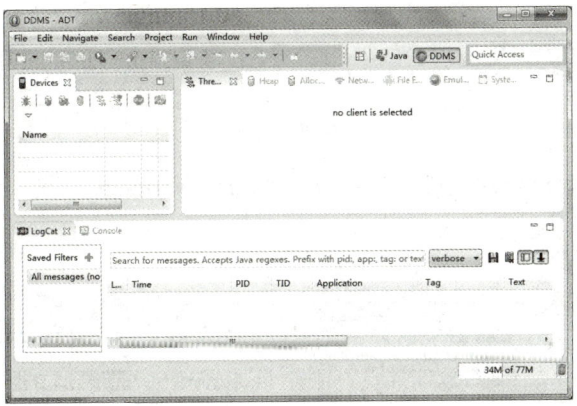

图 4-15　DDMS 模式

如果在使用过程中某些窗口无意关闭，可以通过以下方式打开。

（2）丢失 File Explorer：参照图 4-10 进入 Show View 对话框，打开 File Explorer，如图 4-16 所示。

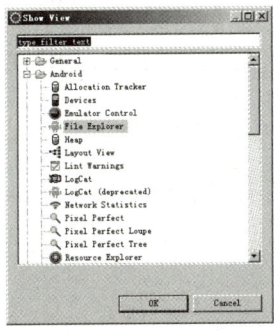

图 4-16　打开 File Explorer

在 File Explorer 中，能看到 Android 开发终端里的目录结构，如图 4-17 所示。

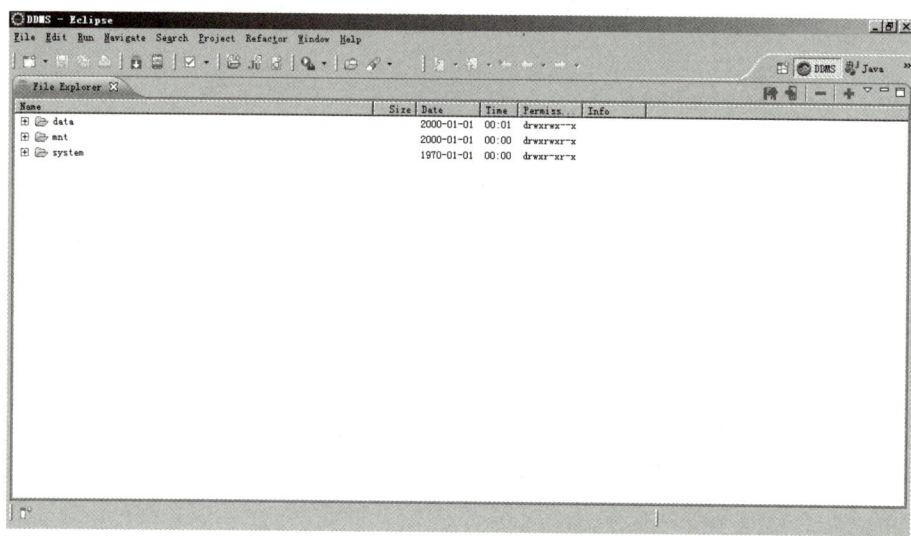

图 4-17　File Explorer 界面

（3）丢失 List Warnings：同样的方法启动 List Warnings 工具，该工具会提示工程存在的错误及警告信息，如图 4-18 所示。

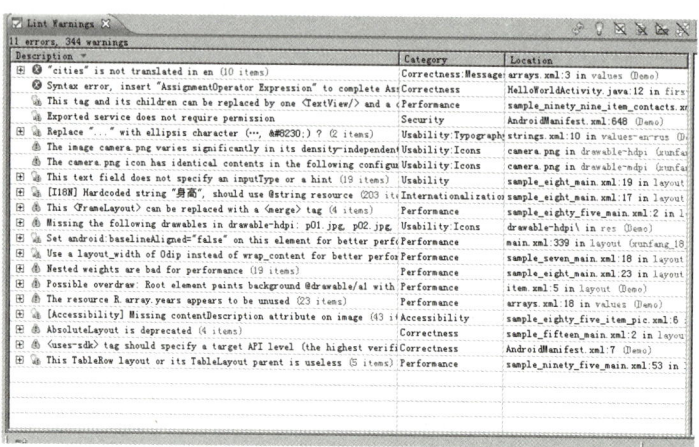

图 4-18　Lint Warings 界面

（4）丢失 LogCat：同法启动，该工具会输出 Android 开发终端的运行信息，如图 4-19 所示。

项目四 Android 工程创建与资源使用

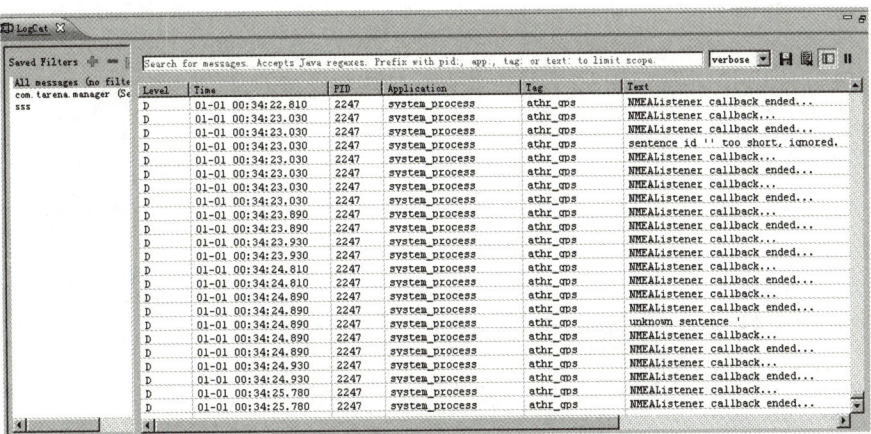

图 4-19　LogCat 界面

四、项目思考与扩展

尝试建立一个项目，输出中文："世界你好！"。

基础篇

项目五　Android 常用界面控件

【本章导读】

　　Android 系统拥有华丽的操作界面,这一点毫不亚于 IOS 系统,这使得 Android 手机深受用户的喜爱。Android 系统中,提供了丰富和华丽的控件,设计者可以实现各具特色的控件和界面效果。本章我们就来学习 Android 系统一些基本组件的使用,完成 Android 基本 UI 的设计。通过这些大量功能丰富的 UI 组件,可以像搭积木一样,开发出优秀的用户界面。

一、项目要求

1. 掌握界面控件 TextView 和 EditText 的使用方法。
2. 掌握界面控件 Button 的使用方法。
3. 掌握界面控件 CheckBox 和 RadioButton 的使用方法。
4. 掌握界面控件 Spinner 的使用方法。

二、项目相关知识

1. Android 视图类

　　Android 中的 View 类是所有 Android 可视化组件的父类,它提供了设置控件和动作事件的函数。在 UI 制作中经常使用的组件如 TextView、Button、CheckBox,都继承于这个类。这些子类控件的相关属性,可以通过在其布局的 XML 文件中进行设置。在程序编写

项目五 Android 常用界面控件

时，这些组件的属性值可以在 XML 程序文件中通过"android:"的形式实现。

下面介绍 Android 视图类组件中通用的一些属性和方法，在后续的学习中我们将会经常用。

1）android:id

该属性定义了视图控件的唯一标识，该标识会以一个类型为十六进制整型常量的形式存放在 R 文件中。编写程序时，一般将 id 的名字作为唯一标识去调用组件。例如：

android:id= "@+id/demo"，表示该组件的标识名为 demo。

2）android:layout_width

该属性定义了视图组件中的宽度属性，其属性值一般可以由以下 3 种形式进行表示：

➢ "match_parent"表示组件的宽度与其父容器一致。
➢ "fill_parent"在 Android 2.2 以后的版本中就不再被使用，它的使用方法与"match_parent"一致。
➢ "wrap_content"表示组件的宽度是根据组件中内容的情况适中显示，也可以理解为其将组件内容进行包裹。

3）android:layout_heigh

该属性定义了组件的高度，其使用方法与 android:layout_width 一致。

4）android:background

该属性定义了组件的背景设置，一般可以设置背景颜色或背景图片。

5）void findViewById（int id）

该函数的功能是根据给定的组件 id 号找到视图中的对象，对象的类型一般为组件的类型。

2. 本文框 TextView

TextView 的功能是在屏幕上显示文本内容。TextView 组件类似于 Java 中 JLable 组件，Android 中的 TextView 组件可以显示单行文本，也可以显示多行文本，还可以显示带图像的文本。一般通过在 XML 布局文件中使用<TextView>标记添加，其基本的语法格式如下：

```
<TextView
属性列表
>
</TextView>
```

TextView 常用的 XML 属性如表 5-1 所示。

表 5-1　TexView 支持的 XML 属性

XML 属性	描　　述
Android: autoLink	用于指定是否将指定格式的文本转换为可单击的超链接形式，其属性值有 none、web、email、phone、map 和 all
Android:drawableBottom	用于在文本框内文本的底端绘制指定图像，该图像可以是放在 res\drawable 目录下的图片，通过"@drawable/文件名（不包括文件的扩展名）"设置
Android:drawableLeft	用于在文本框内文本的左侧绘制指定图像，该图像可以是放在 res\drawable 目录下的图片，通过"@drawable/文件名（不包括文件的扩展名）"设置
Android: drawableRight	用于在文本框内文本的右侧绘制指定图像，该图像可以是放在 res\drawable 目录下的图片，通过"@drawable/文件名（不包括文件的扩展名）"设置
Android: drawableTop	用于在文本框内文本的顶端绘制指定图像，该图像可以是放在 res\drawable 目录下的图片，通过"@drawable/文件名（不包括文件的扩展名）"设置
Android: gravity	用于设置文本框内文本的对齐方式，可选值有 top、bottom、left、right、center_vertical、fill_vertical、center_horizontal、center、fill、clip_vertical 和 clip_horizontal 等。这些属性值也可以同时指定，各属性值之间用竖线隔开。例如，要制定组件靠右下角对齐，可以使用属性值 right	bottom
Android: hint	用于设置当文本框中文本内容为空时，默认显示的提示文本
Android: inputType	用于指定当前文本框显示内容的文本类型，其可选值有 textPassword、textEmailAddress、phone 和 data 等，可以同时指定多个，使用"|"分隔
Android: singleLine	用于指定该文本中显示的文本内容，其属性值为 true 或 false，为 true 表示该文本框不会换行，当文本框中的文本超过一行时，其超出部分将被省略，同时在结尾处添加"……"
Android: text	用于指定该文本中显示的文本内容，可以直接在该属性值中指定，也可以通过在 strings.xml 文件中定义文本常量的方式指定
Android: textColor	用于设置文本框内文本的颜色，其属性值可以通过#rgb、#argb、#rrggbb 或#aarrggbb 格式指定颜色值
Android: textSize	用于设置文本框内文本的字体大小、其属性值由代表大小的数值和单位组成，其单位可以是 px、pt、sp 和 in 等
Android: width	用于指定文本的宽度，以像素为单位
Android: height	用于指定文本的高度，以像素为单位

3. 按钮

程序设计语言中最为常见的事件和动作处理组件就是按钮，Android 中将按钮组件定义为 Button，其提供了普通按钮和带图片按钮两种类型的按钮。这两种按钮都可以通过设置单击事件的方式去实现动作监听。下面对这两种常见的按钮形式进行简单介绍。

1）普通按钮

通过 XML 布局文件的形式进行按钮设置，语法格式如下：

<Button

属性列表>

</Button>

按钮的属性使用方法与 TextView 类似，比较常见的两个属性如下：

（1）android:text 表示按钮显示其文本。

（2）android:background 表示设置按钮的背景颜色。

2）图片按钮

图片按钮与普通按钮的使用方法基本是相同的，在定义的时候将其定义为：

<ImageButton>

</ImageButton>

图片按钮在使用的时候经常应用到 android:src 这个属性来设置按钮的图片来源，图片来自于目录结构中的 drawable 文件夹。

4. 单选按钮和复选框按钮

单选按钮（RadioButton）和复选框按钮（CheckBox）是开发 UI 界面时常用到的基本控件，它们都继承于 Android 中 Button 类，所以它们可以使用 Button 类中所含有的控件属性和方法。

RadioButton 和 CheckButton 与一般的 Button 按钮又有不同之处，它们都多了一个可选中的属性，这个属性可以通过 android:checked 属性来实现，该属性的作用是用于确定 RadioButton 和 CheckButton 在初始化的时候是否被确定选中。

RadioButton 为一个圆形框选择按钮，RadioButton 每次只能选择其中的一个选项，所以 RadioButton 经常与 RadioGroup 一起使用，用于定义一组单选按钮。

RadioGroup 是一个可以容纳多个 RadioButton 组件的容器。

CheckBox 是一个方形框选择按钮，与 RadioButton 不同的是，在同一个时刻可以有多个 CheckBox 处于被选的状态。

通过 XML 布局文件设置单选按钮和复选框按钮的基本语法格式如下：

单选按钮语法格式：

<RadioButton

属性列表>

</RadioButton>

复选框语法格式：

<CheckButton

属性列表>

</CheckButton>

5. 列表选择框

在 Android 中，为我们提供了类似于 JSP 网站开发中常见的下拉框列表，这种列表称为 Spinner 组件，在该组件中提供一系列的列表选择项供用户选择，方便用户的使用。

在使用 Spinner 时，应该注意以下两个属性：

（1）android:entries 表示可选属性，用于指定列表项。

（2）android:prompt 属性也是可选属性，用于指定列表选项的标题。

设置列表选择框的基本语法格式如下：

<Spinner

属性列表>

</Spinner>

三、项目实施过程

下面应用项目二中介绍的 TextView、EditText、RadioButton、Spinner、CheckBox、Button 等 android 基本控件的相关知识，设计如图 5-1 所示的用户注册界面。

项目五　Android 常用界面控件

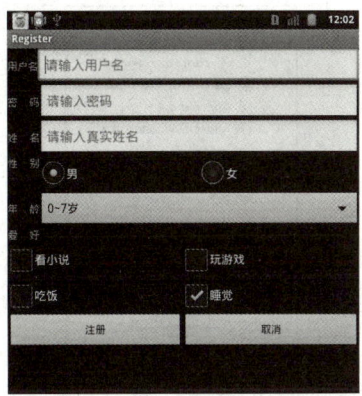

图 5-1　项目效果图

1. 项目创建

在 Android 工程中创建名为 AndroidCode05 的 Android 工程，包结构名为"com.xdxy.register"，Activity 名为 MainActivity，如图 5-2 所示。

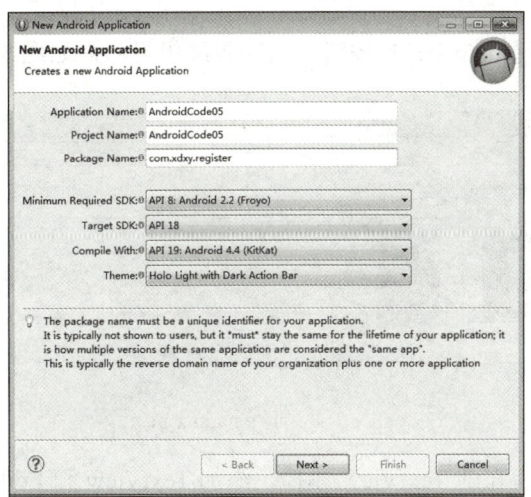

图 5-2　创建工程界面

2. XML 布局文件的开发

1）通过【res】→【layout】路径，找到 main.xml 文件，双击打开，如图 5-3 所示。

57

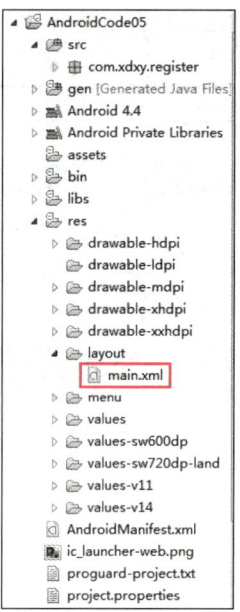

图 5-3　工程目录

2）单击状态栏的【main.xml】按钮进入代码编辑界面，如图 5-4 所示。

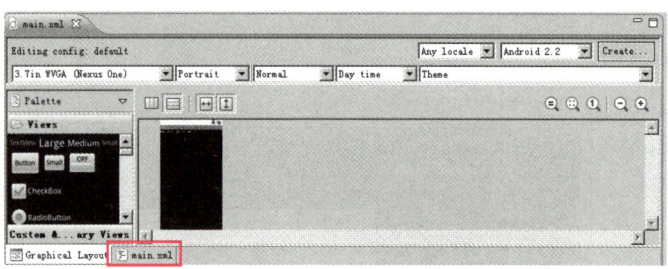

图 5-4　打开 main.xml

3）将原来界面上存在的 TextView 删掉，添加 TextView 和 EditText，一共 3 组，分别输入"用户名""密码"和"姓名"，具体代码如下：

```
<LinearLayout xmlns:android="http://schemas.android.com/apk/res/android"
    android:layout_width="fill_parent"
    android:layout_height="fill_parent"
    android:orientation="vertical">
    <LinearLayout
        android:layout_width="fill_parent"
        android:layout_height="wrap_content" >
```

```xml
        <TextView
            android:layout_width="wrap_content"
            android:layout_height="wrap_content"
            android:text="用户名" />
        <EditText
            android:id="@+id/etUserName"
            android:layout_width="0dp"
            android:layout_height="wrap_content"
            android:layout_weight="1.0"
            android:hint="请输入用户名" />
    </LinearLayout>
    <LinearLayout
        android:layout_width="fill_parent"
        android:layout_height="wrap_content" >
        <TextView
            android:layout_width="wrap_content"
            android:layout_height="wrap_content"
            android:text="密码" />
        <EditText
            android:id="@+id/etUserPass"
            android:layout_width="0dp"
            android:layout_height="wrap_content"
            android:layout_weight="1.0"
            android:hint="请输入密码"
            android:inputType="textPassword" />
    </LinearLayout>
    <LinearLayout
        android:layout_width="fill_parent"
        android:layout_height="wrap_content" >
        <TextView
            android:layout_width="wrap_content"
            android:layout_height="wrap_content"
            android:text="姓名" />
```

```
    <EditText
        android:id="@+id/etRealName"
        android:layout_width="0dp"
        android:layout_height="wrap_content"
        android:layout_weight="1.0"
        android:hint="请输入真实姓名" />
</LinearLayout>
```

4）添加单选框，用于选择性别。先使用 RadioGroup，RadioGroup 里面添加两个 RadioButton 单选按钮，分别为"男"和"女"，具体代码如下：

```
<LinearLayout
    android:layout_width="fill_parent"
    android:layout_height="wrap_content" >
    <TextView
        android:layout_width="wrap_content"
        android:layout_height="wrap_content"
        android:text="性别" />
    <RadioGroup
        android:id="@+id/rgSex"
        android:layout_width="0dp"
        android:layout_height="wrap_content"
        android:layout_weight="1.0"
        android:orientation="horizontal"
        android:checkedButton="@+id/rdoMale">
        <RadioButton
            android:id="@id/rdoMale"
            android:layout_width="0dp"
            android:layout_height="wrap_content"
            android:layout_weight="1.0"
            android:text="男" />
        <RadioButton
            android:id="@+id/rdoFemale"
            android:layout_width="0dp"
            android:layout_height="wrap_content"
```

```
            android:layout_weight="1.0"
            android:text="女" />
    </RadioGroup>
</LinearLayout>
```

5）添加 Spinner，用于选择"年龄段"，具体代码如下所示：

```
<LinearLayout
    android:layout_width="fill_parent"
    android:layout_height="wrap_content" >
    <TextView
        android:layout_width="wrap_content"
        android:layout_height="wrap_content"
        android:text="年龄" />
    <Spinner
        android:id="@+id/spRealAge"
        android:layout_width="0dp"
        android:layout_height="wrap_content"
        android:layout_weight="1.0"/>
</LinearLayout>
```

6）添加"爱好"选择框，输出提示文本"爱好"，具体代码如下所示：

```
<LinearLayout
    android:layout_width="fill_parent"
    android:layout_height="wrap_content" >
    <TextView
        android:layout_width="wrap_content"
        android:layout_height="wrap_content"
        android:text="爱好" />
</LinearLayout>
```

7）使用 CheckBox 显示看小说、玩游戏两个爱好，具体代码如下所示：

```
<LinearLayout
    android:layout_width="fill_parent"
    android:layout_height="wrap_content" >
    <CheckBox
        android:id="@+id/chkRead"
```

```
            android:layout_width="0dp"
            android:layout_height="wrap_content"
            android:layout_weight="1.0"
            android:text="看小说" />
        <CheckBox
            android:id="@+id/chkPlayGame"
            android:layout_width="0dp"
            android:layout_height="wrap_content"
            android:layout_weight="1.0"
            android:text="玩游戏" />
</LinearLayout>
```

8) 使用 CheckBox 显示吃饭、睡觉两个爱好，具体代码如下所示：

```
<LinearLayout
    android:layout_width="fill_parent"
    android:layout_height="wrap_content" >
    <CheckBox
        android:id="@+id/chkEat"
        android:layout_width="0dp"
        android:layout_height="wrap_content"
        android:layout_weight="1.0"
        android:text="吃饭" />
    <CheckBox
        android:id="@+id/chkSleep"
        android:layout_width="0dp"
        android:layout_height="wrap_content"
        android:layout_weight="1.0"
        android:checked="true"
        android:text="睡觉" />
</LinearLayout>
```

9) 为界面添加两个 Button 控件，一个用于提交注册，一个用于取消注册。

```
<LinearLayout
    android:layout_width="fill_parent"
    android:layout_height="wrap_content" >
```

```
<Button
    android:id="@+id/btnRegister"
    android:layout_width="0dp"
    android:layout_height="wrap_content"
    android:layout_weight="1.0"
    android:onClick="doClick"
    android:text="注册" />
<Button
    android:id="@+id/btnCancel"
    android:layout_width="0dp"
    android:layout_height="wrap_content"
    android:layout_weight="1.0"
    android:onClick="doClick"
    android:text="取消" />
</LinearLayout>
```

四、项目思考与扩展

根据资料包中提供的源代码，修改项目三中程序：

（1）在程序中加入一个 TextView，用来显示软件操作说明。

（2）在程序中加入几个 CheckBox 用来增加爱好的选项。

（3）增加一个注册项内容为 Email。

修改后的效果如图 5-5 所示。

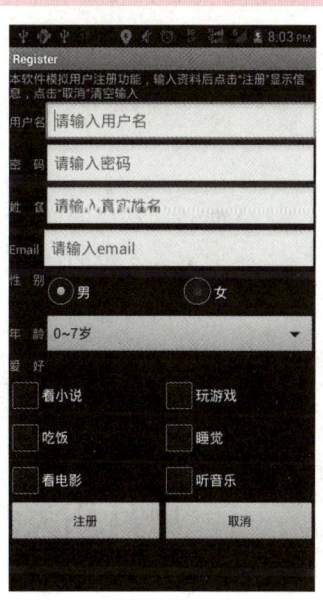

图 5-5　扩展练习程序运行界面

项目六　Android 常用界面布局

【本章导读】

在 Android 应用程序界面中，每个组件在界面中都有合适的位置和具体的大小，默认情况下窗体很难自行对组件的位置进行排列和判断。其实在 Android 中为我们提供了丰富的布局管理器，开发中我们可以使用 Android 的布局管理器对组件的位置和大小进行排列。本项目我们就来学习 Android 中的 4 种布局管理器。

一、项目要求

1. 掌握线性布局的使用。
2. 掌握表格布局的使用。
3. 掌握相对布局的使用。
4. 掌握帧布局的使用。
5. 完成登录界面的开发。

二、项目相关知识

Android 为我们提供了 5 种布局管理器，分别是线性布局管理器（LinearLayout）、表格布局管理器（TableLayout）、帧布局管理器（FrameLayout）、相对布局管理器（RelativeLayout）和绝对布局管理器（AbsoluteLayout）。在 Android 2.2 以后的 Android 版本中，已经把绝对布局管理器（AbsoluteLayout）去掉，所以在本章的项目学习中我们只对前面 4 种布局进行学习。

1. 线性布局

线性布局是将放入其中的组件按照垂直或水平方向来进行布局，也就是说控件是按照水平方向排列或是按照垂直方向排列。在线性布局中，每一行（垂直排列）或是每一列（水平排列）中只能排放一个组件，并且不会换行，当组件排满边缘后，后面的组件将不会再被显示出来。

在 Android 中，一般使用 XML 布局文件对线性布局进行定义，在定义的时候使用 <LinearLayout>对布局进行标记，其基本的语法格式为：

```
<LinearLayout
属性列表
>
</LinearLayout>
```

在线性布局管理器中，经常使用如下属性：

1）android:orientation 属性

android:orientation 属性用于设置布局管理器内组件的排列方式，其可选值为 horizontal 和 vertical，默认情况下为 vertical。horizontal 表示组件水平排列，vertical 表示组件垂直排列。

2）android:gravity 属性

android:gravity 表示线性布局管理器内组件的对齐方式，其可选属性值有 top、bottom、left、right、center_vertical、fill_vertical、center 等。

3）android:layout_width 属性

android:layout_width 属性用于设置组件的基本宽度，其可选数值包括 fill_parent、match_parent 和 wrap_content。其中，fill_parent 和 match_parent 表示组件的宽度与父容器一样，wrap_content 表示组件的宽度恰好可以包裹其显示内容。

4）android:layout_height 属性

android:layout_height 属性用于设置组件的基本高度，其可选数值包括 fill_parent、match_parent 和 wrap_content。其中，fill_parent 和 match_parent 表示组件的高度与父容器一样，wrap_content 表示组件的高度恰好可以包裹好其内容。

5）android:id 属性

android:id 属性用于为当前组件指定一个 id 编号，该属性作为布局管理器和组件的唯一标识，在程序编写时非常重要。

在 XML 文件定义 id 的同时，R.Java 文件中，会自动生成一个对应的十六进制的

整型常量，我们在 Java 程序中可以通过使用 findViewById()这个方法来操作布局管理器和组件。

6）android:background 属性

Android:background 属性用于为组件进行背景的设置，背景可以设置为图片或是颜色，需要注意的是，在为布局管理器设置图片时，必须将背景图片复制到 Android 的图片目录下。使用方式如下：

android:background="@drawable/background"

如果需要将颜色指定为背景的话，可以使用颜色的十六进制数，作为属性数值。例如，需要将背景设置为白颜色，程序可写为：

android:background="#FFFFFFFF"

下面我们来看一个具体的线性布局的 XML 实例。

XML 程序如下：

```
<LinearLayout xmlns:android="http://schemas.android.com/apk/res/android"
    android:layout_width="fill_parent"
    android:layout_height="fill_parent"
    android:gravity="center_horizontal"          //线性布局对齐方式
    android:orientation="vertical" >             //线性布局布局方向
    <Button
        android:id="@+id/btn1"
        android:layout_width="wrap_content"
        android:layout_height="wrap_content"
        android:text="@string/btn1" />
    <Button
        android:id="@+id/btn2"
        android:layout_width="wrap_content"
        android:layout_height="wrap_content"
        android:text="@string/btn2" />
    <Button
        android:id="@+id/btn3"
        android:layout_width="wrap_content"
        android:layout_height="wrap_content"
        android:text="@string/btn3" />
    <Button
```

```
            android:id="@+id/btn4"
            android:layout_width="wrap_content"
            android:layout_height="wrap_content"
            android:text="@string/btn4" />
    <Button
            android:id="@+id/btn5"
            android:layout_width="wrap_content"
            android:layout_height="wrap_content"
            android:text="@string/btn5" />
</LinearLayout>
```

运行效果如图6-1所示。从效果图上我们可以看到，在界面中的所有按钮组件都是以线性垂直的方式在界面中排列。如果将程序中的 android:orientation 属性的值改为 horizontal，我们将会看到这些按钮组件将会以水平的方式进行排列，如图6-2所示。

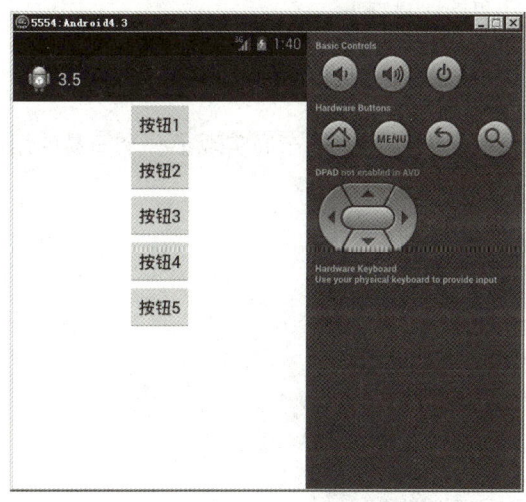
图 6-1　线性布局垂直方向的效果

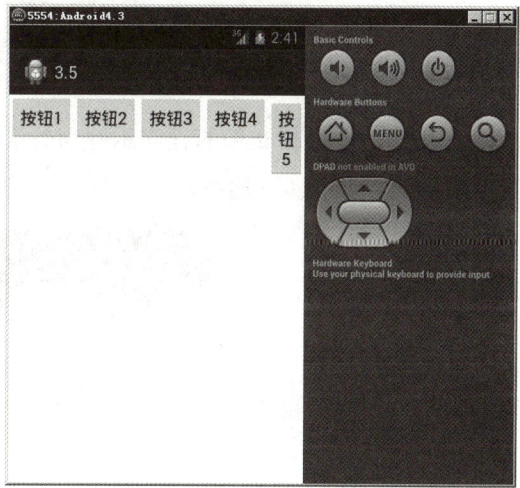
图 6-2　线性布局水平方向的效果

2. 表格布局 TableLayout

Android 程序中表格布局与 Java 等其他程序设计语言中的表格布局相似，都是以行和列的形式存在布局管理器中。在布局管理器的 XML 文件中，使用<TableLayout>的方式对表格布局进行定义，其基本的语法格式为：

```
<TableLayout
    属性列表
```

>
</TableLayout>

在每个表格布局中，可以插入多个表格行，每个表格行单独占据一行，可以使用<TableRow></TableRow>来进行标记。在每个<TableRow>中，可以添加多个组件，每个组件会占据单独的一列。在表格布局中，列可以被隐藏，也可以被伸展，还可以被收缩，这样可以充分利用表格的空间。

TableLayout是LinearLayout的子类，因此TableLayout支持全部的LinearLayout的XML属性，此外TableLayout还支持一些特殊的属性，如表6-1所示。

表6-1　TableLayout支持的XML属性

XML 属性	描　　述
Android：collapseColumns	设置需要被隐藏的列的序列号（序号从0开始），多个序列号之间用逗号","分隔
Android：shrinkColumns	设置允许被收缩的列的序列号（序号从0开始），多个序列号之间用逗号","分隔
Android：stretchColumns	设置允许被拉伸的列的序列号（序号从0开始），多个序列号之间用逗号","分隔

下面是一个表格布局的程序示例和效果图，该示例通过表格布局实现了一个简易的用户登录界面，如图6-3所示。

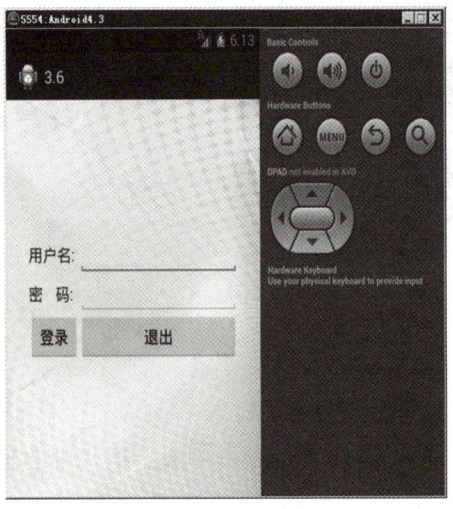

图6-3　表格布局效果图

在布局管理器中添加了3个表格行TableRow，在每个表格行中添加登录界面相关的组件，并将表格的第1列和第4列设置为允许被拉伸。

实例代码如下：

```xml
<TableLayout xmlns:android="http://schemas.android.com/apk/res/android"
    android:id="@+id/tableLayout"
    android:layout_width="fill_parent"
    android:layout_height="fill_parent"
    android:background="@drawable/background"
    android:gravity="center_vertical"
    android:stretchColumns="0,3" >
    <!-- 第 1 行:共列 -->
    <TableRow
    android:id="@+id/tableRow1"
    android:layout_width="wrap_content"
    android:layout_height="wrap_content" >
    <!-- 第 1 列 -->
    <TextView />
    <!-- 第 2 列 -->
    <TextView
    android:id="@+id/textView1"
    android:layout_width="wrap_content"
    android:layout_height="wrap_content"
    android:text="用户名:"
    android:textSize="24px" />
    <!-- 第 3 列 -->
    <EditText
    android:id="@+id/editText1"
    android:layout_width="wrap_content"
    android:layout_height="wrap_content"
    android:minWidth="200px"
    android:textSize="24px" />
    <!-- 第 4 列 -->
    <TextView />
    </TableRow>
    <!-- 第 2 行:共列 -->
```

```xml
<TableRow
    android:id="@+id/tableRow2"
    android:layout_width="wrap_content"
    android:layout_height="wrap_content" >
    <!-- 第 1 列 -->
    <TextView />
    <!-- 第 2 列 -->
    <TextView
        android:id="@+id/textView2"
        android:layout_width="wrap_content"
        android:layout_height="wrap_content"
        android:text="密码:"
        android:textSize="24px" />
    <!-- 第 3 列 -->
    <EditText
        android:id="@+id/editText2"
        android:layout_width="wrap_content"
        android:layout_height="wrap_content"
        android:inputType="textPassword"
        android:minWidth="200px"
        android:textSize="24px" />
    <!-- 第 4 列 -->
    <TextView />
</TableRow>
<!-- 第 3 行:共列 -->
<TableRow
    android:id="@+id/tableRow3"
    android:layout_width="wrap_content"
    android:layout_height="wrap_content" >
    <!-- 第 1 列 -->
    <TextView />
    <!-- 第 2 列 -->
    <Button
```

```
        android:id="@+id/button1"
        android:layout_width="wrap_content"
        android:layout_height="wrap_content"
        android:text="登录" />
        <Button
        android:id="@+id/button2"
        android:layout_width="wrap_content"
        android:layout_height="wrap_content"
        android:text="退出" />
        <!-- 第 4 列 -->
        <TextView />
    </TableRow>
</TableLayout>
```

3. 帧布局

在 Android 应用程序我们每添加一个组件，都会产生一个空白空间，将这个空间称作为一个帧，这些帧将以叠放的形式进行排列，后面的组件将会覆盖前面的组件。定义帧布局管理器的基本语法格式为：

```
<FrameLayout
基本属性>

</FrameLayout>
```

帧布局中主要通过 gravity 属性进行组件的对齐，此外，FrameLayout 支持的 XML 属性及其相关方法如表 6-2 所示。

表 6-2 FrameLayout 支持的 XML 属性及其相关方法

XML 属性	相关方法	说　　明
android:foreground	setForeground (Drawable)	设置该帧布局管理器的前景图像
android:foregroundGravity	setForegroundGravity (int)	设置绘制前景图像的对齐属性

下面是一个帧布局的效果图，该示例将界面各层分别设置为红、橙、黄 3 色，其中中心最上层放置一个图标，如图 6-4 所示。

图 6-4　表格布局效果图

程序如下所示：

```
<FrameLayout xmlns:android=http://schemas.android.com/apk/res/android          //帧布局
    android:id="@+id/frameLayout"
    android:layout_width="fill_parent"
    android:layout_height="fill_parent"
    android:background="@drawable/background"    //帧布局背景图片
    android:foreground="@drawable/icon"          //帧布局前景图标
    android:foregroundGravity="center" >         //帧布局前景对齐方式

<!-- 第 1 层：添加居中显示的红色背景的 TextVIew，最下层 -->
<TextView
    android:id="@+id/textView1"
    android:layout_width="200dp"                 //文本视图组件的宽度：具体值
    android:layout_height="200dp"                //文本视图组件的高度：具体值
    android:layout_gravity="center"
    android:background="#ffff0000" />            //文本视图组件背景色：红色

<!-- 第 2 层：添加居中显示的橙色背景的 TextVIew，中间层 -->
<TextView
    android:id="@+id/textView2"
    android:layout_width="150dp"
    android:layout_height="150dp"
```

```
android:layout_gravity="center"
android:background="#ffff7700"/>        //文本视图组件背景色：橙色

<!-- 第 3 层：添加居中显示的黄色背景的 TextVIew，上层 -->
<TextView
android:id="@+id/textView3"
android:layout_width="100dp"
android:layout_height="100dp"
android:layout_gravity="center"
android:background="#ffffee00" />       //文本视图组件背景色：黄色
```

`</FrameLayout>`

4. 相对布局 RelativeLayout

相对布局是指组件按照其相对位置在布局管理器中进行排列，一般都是以一个组件作为参照物，来确定其他组件的位置，如某个组件在另一个参照组件的上方、下方、左边或右边。相对布局内组件的位置是由兄弟组件和父容器来决定的。如果 A 组件的位置是由 B 组件的位置来决定的，Android 要求先定义 B 组件，再定义 A 组件。

在 Android 的 XML 文件中采用<RelativeLayout></RelativeLayout>的形式来对相对布局文件进行标记。其基本的语法格式为：

```
<RelativeLayout
属性列表>

</RelativeLayout>
```

其主要属性和方法如表 6-3 所示。

表 6-3　RelativeLayouts 支持的 XML 属性及其相关方法

XML 属性	相关方法	说　明
android:gravity	setGravity (int)	设置布局管理器内部各组件的对齐方式
android:ignoreGravity	setIgnoreGravity (int)	设置哪个组件不受 gravity 的影响

RelativeLayout 提供了一个内部类 RelativeLayout.LayoutParams 来控制该布局容器中各子组件的布局分布。RelativeLayout.LayoutParams 提供了大量的 XML 属性来控制 RelativeLayout 布局容器中子组件的布局分布，其主要属性如表 6-4 所示。

表 6-4　RelativeLayout.LayoutParams 支持的 XML 属性

XML 属性	说　明
android:layout_centerHorizontal	控制该子组件是否位于布局容器的水平居中位置
android:layout_centerVertical	控制该子组件是否位于布局容器的垂直居中位置
android:layout_centerInParent	控制该子组件是否位于布局容器的中央位置
android:layout_aliginParentBottom	控制该子组件是否与布局容器底端对齐
android:layout_aliginParentLeft	控制该子组件是否与布局容器左端对齐
android:layout_aliginParentRight	控制该子组件是否与布局容器右端对齐
android:layout_aliginParentTop	控制该子组件是否与布局容器顶端对齐
android:layout_toRightOf	控制该子组件位于给出 ID 组件的右侧
android:layout_toLeftOf	控制该子组件位于给出 ID 组件的左侧
android:layout_above	控制该子组件位于给出 ID 组件的上方
android:layout_below	控制该子组件位于给出 ID 组件的下方
android:layout_alignTop	控制该子组件位于给出 ID 组件的上边界对齐
android:layout_alignBottom	控制该子组件位于给出 ID 组件的下边界对齐
android:layout_alignLeft	控制该子组件位于给出 ID 组件的左边界对齐
android:layout_alignRight	控制该子组件位于给出 ID 组件的右边界对齐

下面是一个相对布局的示例，如图 6-5 所示。从该程序和效果图中，我们可以清楚地看到各个组件之间的位置关系。

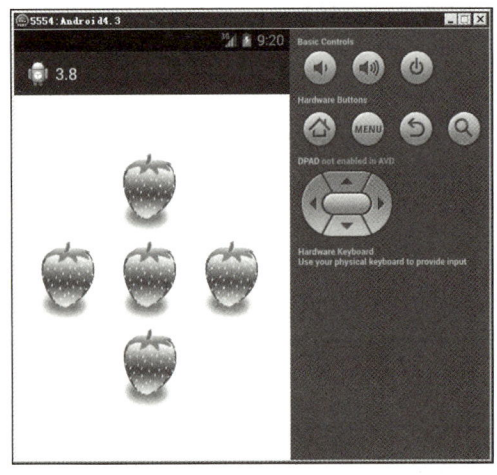

图 6-5　相对布局的效果图

程序示例如下：

<RelativeLayout xmlns:android="http://schemas.android.com/apk/res/android"

```xml
android:layout_width="fill_parent"
android:layout_height="fill_parent" >
<!-- 定义该组件位于父容器中间 -->
<TextView
android:id="@+id/view01"
android:layout_width="wrap_content"
android:layout_height="wrap_content"
android:layout_centerInParent="true"          //控制 view01 组件放在屏幕中央
android:background="@drawable/katong" />
<!-- 定义位于 view01 上方的组件 -->
<TextView
android:id="@+id/view02"
android:layout_width="wrap_content"
android:layout_height="wrap_content"
android:layout_above="@id/view01"    //控制 view02 组件放在指定 view01 组件的上方
android:layout_alignLeft="@id/view01"//控制 view02 组件与指定 view01 组件的左对齐
android:background="@drawable/katong" />

<!-- 定义位于 view01 左方的组件 -->
<TextView
android:id="@+id/view03"
android:layout_width="wrap_content"
android:layout_height="wrap_content"
android:layout_alignTop="@id/view01" //控制 view03 组件与指定 view01 组件的顶端对齐
android:layout_toLeftOf="@id/view01"//控制 view03 组件在指定 view01 组件的左方
android:background="@drawable/katong" />

<!-- 定义位于 view01 右方的组件 -->
<TextView
android:id="@+id/view04"
android:layout_width="wrap_content"
android:layout_height="wrap_content"
android:layout_alignTop="@id/view01"//控制 view04 组件与指定 view01 组件的顶端对齐
```

```
            android:layout_toRightOf="@id/view01"     //控制 view04 组件在指定 view01 组件的右方
            android:background="@drawable/katong" />

        <!-- 定义位于 view01 下方的组件 -->
        <TextView
            android:id="@+id/view05"
            android:layout_width="wrap_content"
            android:layout_height="wrap_content"
            android:layout_alignLeft="@id/view01"//控制 view05 组件与指定 view01 组件的左对齐
            android:layout_below="@id/view01"    //控制 view05 组件在指定 view01 组件的下方
            android:background="@drawable/katong" />
</RelativeLayout>
```

三、项目实施过程

下面我们来做一个简单的登录程序，界面效果如图 6-6 所示。

图 6-6　项目效果图

1. 创建工程

创建名为 AndroidCode06 的 Android 工程，包结构为"com.xdxy.layout"，Activity 名为 MainActivity，如图 6-7 所示。

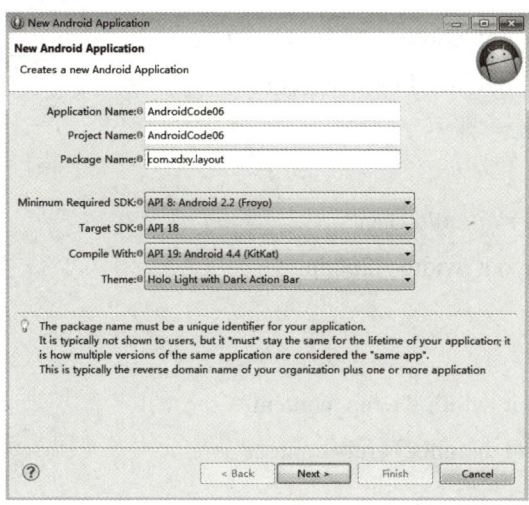

图 6-7　创建工程

2. XML 布局文件的开发

1）通过【res】→【layout】路径，找到 main.xml 文件，双击打开，单击状态栏的【main.xml】按钮进入代码编辑界面。

2）删除 main.xml 自带的 TextView，在线性布局中添加一个 gravity 属性，使位于该线性布局中的组件在水平和垂直方向上居中。

```
<LinearLayout xmlns:android="http://schemas.android.com/apk/res/android"
    android:layout_width="fill_parent"
    android:layout_height="fill_parent"
    android:orientation="vertical"
    android:gravity="center_vertical|center_horizontal">
```

3）在线性布局中添加一个表格布局。

```
<TableLayout
    android:layout_width="wrap_content"
    android:layout_height="wrap_content">
```

4）在表格布局中添加两行，一行用于输入"账号"，一行用于输入"密码"。

```
<TableRow
```

```xml
        android:layout_width="wrap_content"
        android:layout_height="wrap_content" >

        <TextView
            android:layout_width="wrap_content"
            android:layout_height="wrap_content"
            android:text="账号"/>
        <EditText
            android:id="@+id/etUser"
            android:layout_width="200dp" />
    </TableRow>
    <TableRow
        android:layout_width="wrap_content"
        android:layout_height="wrap_content" >
        <TextView
            android:layout_width="wrap_content"
            android:layout_height="wrap_content"
            android:text="密码"/>
        <EditText
            android:id="@+id/etPass"
            android:layout_width="200dp"/>
    </TableRow>
</TableLayout>
```

5）在表格布局下添加一个相对布局。

```xml
<RelativeLayout
    android:layout_width="wrap_content"
    android:layout_height="wrap_content">
```

6）在相对布局里添加两个按钮，分别为登录和退出。

```xml
<Button
    android:id="@+id/btLogin"
    android:layout_width="wrap_content"
    android:layout_height="wrap_content"
    android:onClick="doClick"
```

```
        android:text="登录"/>
<Button
        android:id="@+id/btExit"
        android:layout_toRightOf="@id/btLogin"
        android:layout_width="wrap_content"
        android:layout_height="wrap_content"
        android:onClick="doClick"
        android:text="退出"/>
</RelativeLayout>
```

7）在相对布局下面添加一个帧布局。

```
<FrameLayout
        android:layout_width="150dp"
        android:layout_height="100dp">
```

8）在帧架布局里添加 3 个 TextView。

```
<TextView
        android:id="@+id/tvSuccess"
        android:visibility="invisible"
        android:layout_width="fill_parent"
        android:layout_height="fill_parent"
        android:gravity="center_vertical|center_horizontal"
        android:textColor="#FF00FF00"
        android:text="登录成功"/>
<TextView
        android:id="@+id/tvFail"
        android:visibility="invisible"
        android:layout_width="fill_parent"
        android:layout_height="fill_parent"
        android:gravity="center_vertical|center_horizontal"
        android:textColor="#FFFF0000"
        android:text="登录失败"/>
<TextView
        android:id="@+id/tvGo"
        android:layout_width="fill_parent"
```

```
            android:layout_height="fill_parent"
            android:gravity="center_vertical|center_horizontal"
            android:textColor="#FF0000FF"
            android:text="请登录"/>
    </FrameLayout>
</LinearLayout>
```

9）至此界面设计完成，按【Ctrl+S】组合键保存后关闭 main.xml 文件。

四、项目思考与扩展

1. 修改程序，使账号密码输入框靠右，账号、密码文字靠左。
2. 应用表格布局和线性布局，实现如图 6-8 所示的分类显示快捷工具栏。

图 6-8　显示快捷工具栏效果图

项目七 Android 单击事件的处理

【本章导读】

经过之前项目的学习,我们已经掌握了 Andorid 基本控件和基本布局的知识,这些布局和控件,组成了应用界面。用户在使用手机、平板电脑时,会通过各种操作来与软件进行交互,组件经过触发后需要在应用程序中产生相应的响应。本章我们就来学习如何触发这些控件产生相应动作。

一、项目要求

1. 为界面相应控件添加按键事件监听器及按键事件处理方法。
2. 掌握按钮单击事件的四种方法。
3. 通过 findViewById()方法,将 XML 的组件和 Java 程序进行连接。
4. 完成人民币大小写转换软件的设计与开发。

二、项目相关知识

目前手机端的软件都是图形化界面的软件,大都通过事件机制来实现人机交互,Android 为我们设置了完善的事件监听机制,开发时,我们可以利用这些监听器来实现动作的处理。在 Android 系统中,主要存在两种类型的事件处理:键盘事件和接触事件。本章我们主要学习接触事件的处理。

1. 事件监听原理

Android 中按钮事件监听机制原理与 JAVA 程序中的按钮事件监听机制原理基本一致。如图 7-1 所示，基于监听的事件处理的处理流程如下：

（1）用户按下屏幕中的一个按钮或者单击某个菜单项。

（2）按下动作会激活一个相应的事件，这个事件会触发事件源上注册的事件监听器。

（3）事件监听器会调用对应的事件处理器（事件监听器里的实例方法）来做出相应的响应。

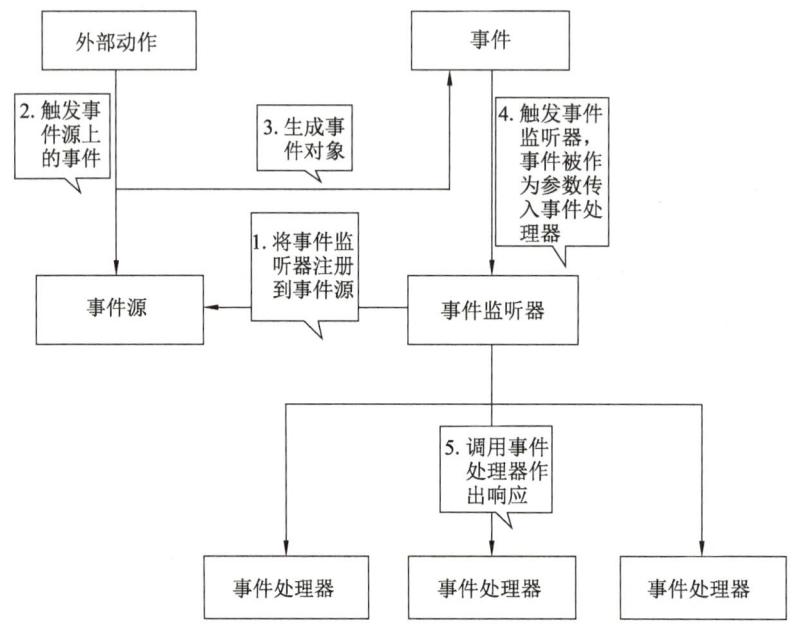

图 7-1　基于监听的事件处理流程

一个组件想要触发动作，必须满足三个部分，分别是事件源、监听器和具体执行的动作，一般将这三个部分称作事件处理机制的三要素。

事件源：事件源就是指事件是由"谁"产生的，就是事件是由哪个组件触发的。

监听器：监听器就是一个实现特定接口的普通 Android 程序，这个程序专门用于监听另一个 Android 对象的方法调用或属性改变，当被监听对象发生上述事件后，监听器某个方法将立即被执行。

在 Android 开发中我们一般可以使用 OnClickListener 和 OnLongClickListener 两种

> 监听器，分别处理用户短时监听和用户长时监听。

具体事件：具体事件就是监听器监听到的具体动作，也就是组件要触发的具体动作。

2. findViewById()方法

Android 的项目框架，是一个典型的 MVC 结构的框架，其 XML 中的组件和 JAVA 程序是分离开的，但在程序开发的过程中，我们需要将二者连接起来，使 JAVA 程序控制组件事件，Android 为我们提供了 findViewById()方法来完成此功能。具体使用方法如下：

下面是在 XML 文件中定义的一个按钮组件：

```
<Button
android:id="@+id/button1"
android:layout_width="wrap_content"
android:layout_height="wrap_content"
android:text="按钮单击" />
```

其中程序 android:id="@+id/button1"该行最为关键，可以理解为组件的 id，也就是组件的唯一标识。该行程序可理解为：@代表 Android 程序框架中的 R.Java 文件，"+"代表增加的意思，id 为组件的唯一标识，button1 为组件的名字。就是说我们通过该行程序，不仅为组件添加了一个名为 button1 的 id，同时在 R.Java 文件中也自动生成了一个名为 button1 的 id。

在 R.Java 文件中我们可以看到这个增加的 id，如下所示：

```
public static final class id {
                public static final int action_settings=0x7f080001;
                public static final int button1=0x7f080000;
        }
```

在 src 文件夹的 MainActivity.Java 文件中我们可以使用如下的程序，进行 XML 中的组件和 JAVA 程序的连接，程序如下：

```
Button but = (Button) findViewById(R.id.button1);
```

在这里我们可以看到 findViewById()方法是连接 XML 组件 button1 和 JAVA 程序中 Button 类型的 but 变量的关键，需要注意的是因为"="号的左边是 Button 类型，"="号的右边是十六进制数，所以我们需要使用转型将左右两边的数据类型进行统一。

3. 按钮单击事件的四种方法

我们这里选择按钮组件，介绍 Android 中的四种程序实现方法。

1）内部类作为事件监听器

```java
btnButton.setOnClickListener(new MyListener());
class MyListener implements OnClickListener {
    public void onClick(View v) {
        System.out.println("内部类响应单击事件");
    }
}
```

2）匿名内部类作为事件监听器

```java
btnButton.setOnClickListener(new OnClickListener() {
        public void onClick(View v) {
            System.out.println("匿名内部类响应按钮单击事件");
        }
});
```

3）自身类作为事件监听器

```java
public class TestButtonActivity extends Activity implements OnClickListener {
    Button btn1, btn2;
    protected void onCreate(Bundle savedInstanceState) {
        super.onCreate(savedInstanceState);
        setContentView(R.layout.activity_test_button);
        btn1 = (Button) findViewById(R.id.button1);
        btn2 = (Button) findViewById(R.id.button2);
        btn1.setOnClickListener(this);
        btn2.setOnClickListener(this);
    }

    public void onClick(View v) {
        switch (v.getId()) {
        case R.id.button1:
            System.out.println("自身类实现 1");
            break;
        case R.id.button2:
            System.out.println("自身类实现 2");
            break;
```

```
                default:
                    break;
                }
            }
        }
```

4）在 XML 文件中"显示指定按钮的 onClick 属性"

XML 程序：

```xml
<Button
    android:id="@+id/button1"
    android:layout_width="wrap_content"
    android:layout_height="wrap_content"
    android:onClick="onClick"
    android:text="Button1" />
<Button
    android:id="@+id/button2"
    android:layout_width="wrap_content"
    android:layout_height="wrap_content"
    android:onClick="onClick"
    android:text="Button2" />
```

Java 程序：

```java
public class TestButtonActivity extends Activity {
    Button btn1, btn2;
    Toast tst;
    protected void onCreate(Bundle savedInstanceState) {
        super.onCreate(savedInstanceState);
        setContentView(R.layout.activity_test_button);
    }
    public void onClick(View v) {
        switch (v.getId()) {
        case R.id.button1:
            tst = Toast.makeText(this, "111111111", Toast.LENGTH_SHORT);
            tst.show();
            break;
```

```
            case R.id.button2:
                tst = Toast.makeText(this, "222222222", Toast.LENGTH_SHORT);
                tst.show();
                break;
        }
    }
}
```

在进行项目开发的过程时,我们可以自行选择其中的任何一种方法进行单击事件的处理。

三、项目实施过程

在本章项目中我们学习了 Android 事件的处理方法,在之前的项目中我们学习了 Android 的基本布局和基本组件,现在我们将之前所学到的内容结合起来,开发一个 APP 程序,该软件可以将小写阿拉伯数字转换为大写人民币,具体效果如图 7-2 所示。

图 7-2　程序界面效果图

1. 创建工程

创建名为 AndroidCode07 的 Android 工程,包结构为 "com.xdxy.rmb",Activity 名为 RmbActivity,如图 7-3 所示。

项目七 Android 单击事件的处理

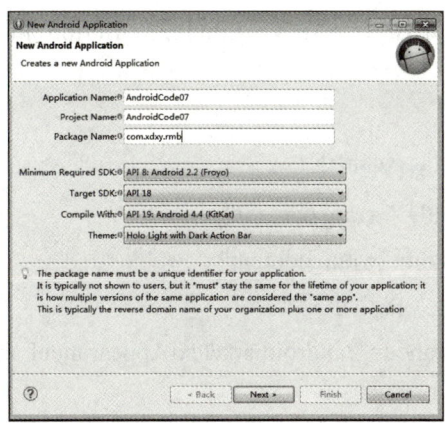

图 7-3　创建工程

2. XML 布局文件的开发

1）通过【res】→【layout】路径，找到 main.xml 文件，双击打开，接着单击状态栏的【main.xml】按钮进入代码编辑界面，如图 7-4 所示。

图 7-4　main.xml 文件

2）删除 main.xml 自带的 TextView，使用线性布局作为界面的布局，线性布局的方向设置为线性垂直方向，并为界面制定一张名为 bj1 的图片，程序如下所示：

```
<LinearLayout xmlns:android="http://schemas.android.com/apk/res/android"
    android:layout_width="fill_parent"
    android:layout_height="fill_parent"
    android:background="@drawable/bj1"
    android:orientation="vertical" >
</LinearLayout>
```

3）在线性布局中依次添加 TextView、EditText、Button 和 TextView 组件，程序如下所示：

```xml
<TextView
    android:id="@+id/textView1"
    android:layout_width="wrap_content"
    android:layout_height="wrap_content"
    android:text="输入人民币小写"
    android:textAppearance="?android:attr/textAppearanceLarge" />
<EditText
    android:id="@+id/editText1"
    android:layout_width="match_parent"
    android:layout_height="wrap_content"
    android:ems="10"
    android:inputType="numberDecimal" >
<requestFocus />
</EditText>
<Button
    android:id="@+id/button1"
    android:layout_width="wrap_content"
    android:layout_height="wrap_content"
    android:text="转换" />
<TextView
    android:id="@+id/textView2"
    android:layout_width="wrap_content"
    android:layout_height="wrap_content"
    android:text="运行结果"
    android:textAppearance="?android:attr/textAppearanceLarge" />
```

3. Java 文件的开发

1）在新建的工程下找到 src 目录，在 src 下创建一个名为 Rmb.Java 的文件，该程序的作用是将小写的阿拉伯数字转换为大写的数字。该程序中的 setrmb(String rmbstr)函数和 getrmb()函数非常重要，在后续的开发中需要调用这两个函数。程序如下：

```java
public class Rmb {
```

```java
String a[]={"零","壹","贰","叁","肆","伍","陆","柒","捌","玖"};
String c[]={"","亿","千","百","拾","万","千","百","拾","亿","千","百","拾","万",
                                                "千","百","拾","元"};
String d[]={"","角","分","厘",""};
String rmb,rmbstrtemp;
String rmbint,rmbdec;
void setrmb(String rmbstr)
{
    rmbstrtemp=rmbstr;
}
String getrmb()
{
    StringBuffer rmbstrResult=new StringBuffer();
    StringBuffer rmbstrall=new StringBuffer();
    if((rmbstrtemp.indexOf(".")>0)&&(rmbstrtemp.length()-rmbstrtemp.indexOf(".")>3))
    //整理输入的数据通过判断输入的字符串是否有小数位
    //而且小数位数大于3位，以确保小数位只有两位；
    {
        rmbstrtemp=rmbstrtemp.substring(0,
            rmbstrtemp.length()-((rmbstrtemp.length()- rmbstrtemp.indexOf("."))-3));
        //12345.123        9-((9-6)-3)    9-(3-3) 9-(0) 9
        //12345.1234       10-((10-6)-3) 10-(4-3) 10-1 9
        //12345.12345      11-((11-6)-3) 11-(5-3) 11-2 9
    }
    rmbstrall.append(rmbstrtemp);
    if(rmbstrtemp.indexOf(".")<0 )
    //判断输入的字符串有没有小数点如果是加".00"；
    {
            rmbstrall.append(".00");
    }
    if((rmbstrtemp.length()-rmbstrtemp.indexOf("."))==1)
    //判断输入的字符串是不是只带小数点不带小数位如果是在后面加"00";
    {
```

```
            rmbstrall.append("00");
        }
        rmb=rmbstrall.toString();
        rmbint=rmb.substring(0,rmb.indexOf("."));              //取出整数部分
        rmbdec=rmb.substring(rmb.indexOf(".")+1,rmb.length()); //取出小数部分
        String rmbchar;
        int rmbdic,rmbcom;
        rmbcom=c.length-(rmbint.length());      //判断整数的单位开始位置
        for(int i=0;i<rmbint.length();i++)
        {
            rmbchar=rmbint.substring(i,i+1);    //取出整数的每一个位的字符从左往右
            rmbdic=Integer.parseInt(rmbchar);   //将每一个位的字符转换为整数
            rmbstrResult.append(a[rmbdic]);
        //将每一个位的整数与a［］数组下标对应的中文大写字符存入 rmbstrResult
            rmbstrResult.append(c[rmbcom]);          //将单位跟在大写字符后面
            rmbcom++;
        }
        for(int i=0;i<rmbdec.length();i++)
        {
            rmbchar=rmbdec.substring(i,i+1);
            rmbdic=Integer.parseInt(rmbchar);
            rmbstrResult.append(a[rmbdic]);
            rmbstrResult.append(d[i+1]);
        }
        return rmbstrResult.toString();
    }
}
```

2）在 src 下打开 RmbActivity.Java，进行转换动作的编写，首先对组件进行声明，程序如下：

```
public class RmbActivity extends Activity {
    TextView text1;
    Button btn1,btn2;
    EditText edit1;
```

项目七　Android 单击事件的处理

```
    Rmb rmb1;
    protected void onCreate(Bundle savedInstanceState) {
        super.onCreate(savedInstanceState);
        setContentView(R.layout.activity_rmb);
        text1=(TextView)findViewById(R.id.textView2);
        btn1=(Button)findViewById(R.id.button1);
        edit1=(EditText)findViewById(R.id.editText1);
```

3）将 Rmb 类进行实例化，方便以后调用：

```
rmb1=new Rmb();
```

4）为转换按钮设置监听动作，这里调用了 Rmb.Java 程序中的 setrmb(String rmbstr)函数和 getrmb()函数，程序如下所示：

```
btn1.setOnClickListener(
new OnClickListener(){
        public void onClick(View v){
                rmb1.setrmb(edit1.getText().toString());
                text1.setText(rmb1.getrmb().toString());
                }
    }
);
```

四、项目思考与扩展

1. 为本章人民币大小写软件程序开发出一个清空按钮。
2. 开发如图 7-5 所示的计算器应用程序。

图 7-5　计算器应用程序

项目八　Intent 实现消息传递

【本章导读】

　　在前面介绍的实例中已经应用过 Activity，不过那些实例中的所有操作都是在一个 Activity 中进行的，在实际的应用开发中，经常需要包含多个 Activity，而且这些 Activity 之间可以相互跳转或传递数据。如果用户想从一个界面跳转另一个界面，需要应用 Intent 进行跳转，在界面跳转之后会将数值传送到另一个界面之中。本章我们就来学习如何使用 Intent 传递消息。

一、项目要求

1. 掌握显式启动 Activity 的方法。
2. 掌握隐式启动 Activity 的方法。
3. 掌握通过 Intent 传递数据的方法。
4. 完成用户注册项目。

二、项目相关知识

1. Intent 概述

　　Intent 是一种轻量级的消息传递机制，用于组件之间数据交换和发送广播消息。它可以在同一应用程序内的不同组件间传递信息，也可以在不同应用程序的组件间传递信息，还可以作为广播事件发布 Android 系统消息。在 Intent 对象中，包含了接收该 Intent 的组件信息和 Android 系统信息，一般来说，一个完整的 Intent 包含组件的名称、动作、数据、种类、额外和标记等一系列内容。下面我们将对这些内容进行具体的介绍。

1）组件名称（Component Name）

组件名称是指 Intent 所在的目标组件的名称。它是一个 ComponeName 类的对象，由完全限定类名（如 com.xdxy.ActivityDemo）和组件所在应用程序配置文件中设置的包名（如 com.xdxy）组合而成。

组件名称的包名部分和配置文件中设置的包名不必匹配，组建名称是可选的。组件名称可以使用 setComponent()、setClassName()方法设置，使用 get Component()方法读取。如果设置，Intent 对象会被发送给指定类的实例；如果没有设置，Android 使用 Intent 对象中的其他信息决定合适的目标。

2）动作（Action）

Action 是一个字符串，用来表示将要执行的动作。在广播 Intent 中，Action 用来表示已经发生即将报告的动作。在 Intent 类中，定义了一系列动作常量,其目标组件包括 Activity 和 Broadcast 两类。

3）数据（Data）

Data 表示操作数据的 URI 和 MIME 类型。不同动作与不同类型的数据规范匹配。

4）种类（Category）

Category 也是一个字符串，其中包含了应该处理的当前 Intent 组件的类型中的特殊信息。

5）额外（Extras）

Extras 是一组键值对，其中含有需要传递给 Intent 的组件的额外信息。

6）标记（Flags）

Flags 表示不同的来源标记，一般所有的标记都被定义在 Intent 类中。

2. 显性 Intent

一般情况下，一个 Android 应用程序中需要多个界面，每个界面就是一个窗口，每个窗口就是一个 Activity，这些 Activity 之间是通过 Intent 的跳转机制来切换的。Intent 在使用的时候分为两种：显性 Intent 和隐性 Intent。在同一个应用程序中切换不同的 Activity 时，我们需要知道具体哪个 Activity 被组件启动或激活，所以常用显性的 Intent 来实现。

下面是一个使用显性 Intent 进行跳转的例子，MainActivity 和 SecondActivity 分别是在同一应用程序中的两个 Activity 文件，现用 MainActivity 跳转到 SecondActivity。在程序中，主要是为【转到 SecondActivity】按钮添加了 OnClickListener，使得按钮被单击时执行 onClick()方法，onClick()方法中则利用 Intent 机制来启动 SecondActivity。具体实现过程如下：

（1）首先编写两个布局 XML 文件，如下所示。
➢ 第一个布局文件，名字为：main.xml。

```xml
<?xml version="1.0" encoding="utf-8"?>
<LinearLayout xmlns:android="http://schemas.android.com/apk/res/android"
    android:orientation="vertical"
    android:layout_width="fill_parent"
    android:layout_height="fill_parent">

    <Button
        android:id="@+id/btn"
        android:layout_width="wrap_content"
        android:layout_height="wrap_content"
        android:text="转到第二个界面"/>

</LinearLayout>
```

➢ 第二个布局文件，名字为：second.xml。

```xml
<?xml version="1.0" encoding="utf-8"?>
<LinearLayout xmlns:android="http://schemas.android.com/apk/res/android"
    android:orientation="vertical"
    android:layout_width="fill_parent"
    android:layout_height="fill_parent">

    <TextView
        android:layout_width="fill_parent"
        android:layout_height="wrap_content"
        android:text="这是第二个界面"/>

</LinearLayout>
```

（2）在 MainActivity.Java 文件中利用 Intent 实现跳转。

```java
public class MainActivity extends Activity {
    private Button btn;
    public void onCreate(Bundle savedInstanceState) {
        super.onCreate(savedInstanceState);
```

```
            setContentView(R.layout.main);
            btn = (Button)findViewById(R.id.btn);
            //响应按钮 btn 事件
            btn.setOnClickListener(new OnClickListener() {
                public void onClick(View v) {
                    //显示方式声明 Intent，直接启动 SecondActivity
                    Intent it = new Intent(MainActivity.this,SecondActivity.class);
                    //启动 Activity
                    startActivity(it);
                }
            });
        }
}
```

在该程序，我们可以看到进行跳转时，只要在 Intent 的实例化构造函数的参数中，将当前界面（MainActivity.this）和需要跳转到的界面（SecondActivity.class）的跳转路径设置清楚后，再使用 startActivity()方法就可以实现界面之间的跳转。

（3）建立 SecondActivity.Java 程序，为了实现跳转，我们需要建立一个对应的 Java 程序。在该程序中，只需要让这个类继承于 Activity 类即可（普通的类只有继承于 Activity 类才能是 Activity 类），具体程序如下：

```
public class SecondActivity extends Activity {
    protected void onCreate(Bundle savedInstanceState) {
        super.onCreate(savedInstanceState);
        setContentView(R.layout.second);
    }
}
```

（4）在 Android 的清单文件 AndroidManifest.xml 中完成对 SecondActivity 的注册。具体程序如下所示：

```
<?xml version="1.0" encoding="utf-8"?>
<manifest xmlns:android="http://schemas.android.com/apk/res/android"
package="com.android.test.activity"
android:versionCode="1"
android:versionName="1.0">
<uses-sdk android:minSdkVersion="10" />
```

```xml
<application android:icon="@drawable/icon" android:label="@string/app_name">
<activity android:name=".MainActivity"
android:label="@string/app_name">
<intent-filter>
<action android:name="android.intent.action.MAIN" />
<category android:name="android.intent.category.LAUNCHER" />
</intent-filter>
</activity>
<activity android:name=".SecondActivity"
android:label="@string/app_name">
</activity>
</application>
</manifest>
```

（5）当单击第一个页面中的【跳转到第二个界面】按钮后，即可跳转至第二个界面，实现效果如图 8-1 所示。

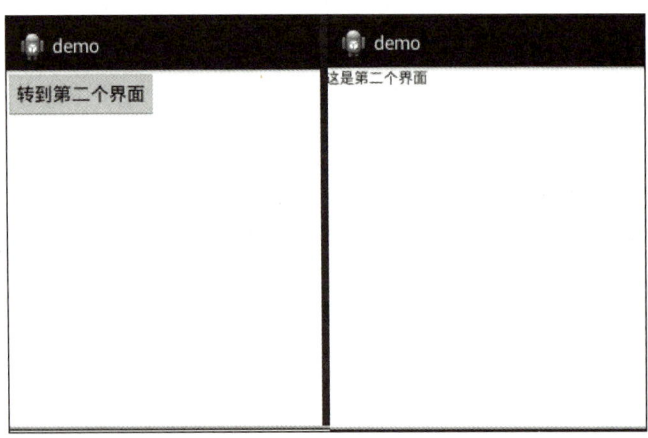

图 8-1　使用显性 Intent 方式进行跳转的效果

3. 显性 Intent 的数据传递

一般在 Activity 的跳转中，如果我们希望把一个 Activity 界面中的某些数据和数值传递给下一个 Activity。这时，可以使用 Bundle 类来实现。在 Android 中的 Bundle 类相当于数据的一个存储包，用于储存我们需要发送和接收的数据。与现实生活中使用的 Email 发送邮件一样，Intent 就像一封邮件，里面有发送邮件人的地址（发送事件的 Activity），也有接收邮件人的地址（接收事件的 Activity），而 Bundle 类就是在邮件中的附件，也就是我们邮件传送的内容。下面是其示例程序：

1）发送数据 ActivityFirst.Java

```
Intent intent = new Intent();
intent.setClass(ActivityFirst.this, ActivitySecond.class);
//设置发送数据 Activity 和接收数据 Activity
//ActivityFirst 为发送数据的 Activity，ActivitySecond.为接收数据的 Activity
Bundle bundle = new Bundle();      //实例化一个 Bundle 对象，对象的名字叫做 bundle
bundle.putString("demo", "ActivityFirst 发来的数据");   //使用 putString()方法装入数据
                                                       //demo 为发送数据名
intent.putExtras(bundle);                //把 Bundle 放入 Intent 里面
startActivity(intent);                   //启动并开始发送数据
```

2）ActivitySecond 接受从 ActivityFirst 发来的数据

```
Intent intent = this.getIntent();        //获取已有的 intent 对象
Bundle bundle = intent.getExtras();      //获取 intent 里面的 bundle 对象
string = bundle.getString("demo");       //获取 Bundle 里面的字符串
```

4. 隐性 Intent

隐性 Intent 主要用于激发 Android 系统出的自带组件，如：拨打电话、上网等功能。下面我们来看一个 Android 隐性 Intent 的例子，该程序是一个电话拨打的应用程序。

1）XML 布局程序

```xml
<RelativeLayout xmlns:android="http://schemas.android.com/apk/res/android"
    xmlns:tools="http://schemas.android.com/tools"
    android:layout_width="match_parent"
    android:layout_height="match_parent"
    tools:context=".MainActivity" >
    <EditText
        android:id="@+id/etPhone"
        android:layout_width="match_parent"
        android:layout_height="wrap_content">
    </EditText>
    <Button
        android:id="@+id/btnCall"
```

```
                android:layout_width="wrap_content"
                android:layout_height="wrap_content"
                android:layout_alignLeft="@+id/etPhone"
                android:layout_below="@+id/etPhone"
                android:text="拨号" />
</RelativeLayout>
```

2）MainActivity 程序

```
public class MainActivity extends Activity {
    protected void onCreate (Bundle savedInstanceState) {
        super.onCreate(savedInstanceState);
        setContentView(R.layout.activity_main);           //设置页面的布局
        Button tel = (Button)findViewById(R.id.btnCall);  //通过 id 值获取用户编辑框对象
        tel.setOnClickListener(new OnClickListener()
        {
            //设置按钮监听事件
            public void onClick(View v) {
                //获得用户输入的电话号码
                String phoneNumber =((EditText) findViewById(R.id.etPhone)).getText().toString();
                Intent intent = new Intent();                          //创建意图
                intent.setAction(Intent.ACTION_CALL);                  //指定拨打电话动作
                intent.setData(Uri.parse("tel:" + phoneNumber));       //进行 URL 统一资源定位
                startActivity(intent);                                 //开启系统拨号器
            }
        });
    }
}
```

在该程序片段中，intent.setAction(Intent.ACTION_CALL)表示设置 Intent 的动作为打电话，intent.setData(Uri.parse("tel:" + phoneNumber))是设置要传递的数据，需要注意的是在 Uri 中一定要在电话号码要加前缀"tel:"。

3）在 Android 清单文件 AndroidManifest.xml 中注册动作

```
<uses-sdk
    android:minSdkVersion="8"
    android:targetSdkVersion="19" />
```

```
<uses-permission android:name="android.permission.CALL_PHONE"
                                    android:maxSdkVersion="19"/>
```

对拨打电话的权限进行注册后，才能保证程序的正常运行。

三、项目实施过程

下面我们通过做一个个人信息录取程序，来学习 Intent 的实际开发操作。程序界面如图 8-2 所示。

（a）
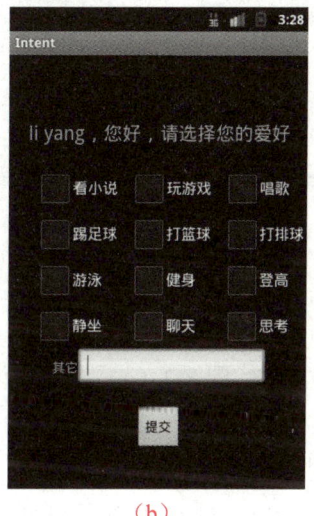
（b）
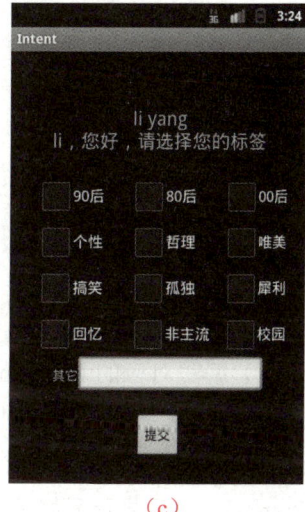
（c）

图 8-2　项目效果图

1. 创建工程

创建名为 AndroidCode08 的 Android 工程，包结构为"com.xdxy.intent"，创建的 Activity 名为 IntentActivity，如图 8-3 所示。

Android 移动开发项目化教程

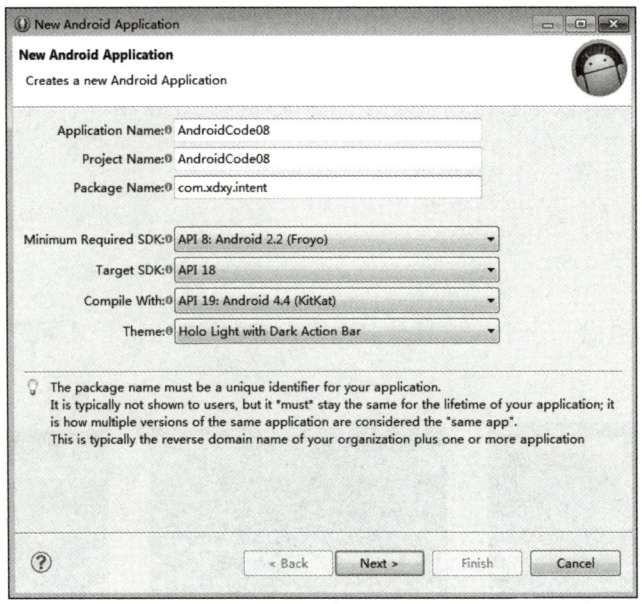

图 8-3　创建工程

2. XML 布局文件的开发

（1）编写对应的 main.xml 文件，图 8-2（a）界面整体是一个垂直的线性布局，该布局内部控件在水平及垂直方向上均居中显示。其内部主要控件从上到下依次是一个用于显示界面主题的 TextView，一个用于输入姓名的文本编辑框 EditText，用于显式跳转的按钮 1（Button），用于显示爱好信息的 TextView（默认不可见），用于显式跳转的按钮 2（默认不可见），用于隐式跳转的按钮 1，用于显示标签信息的 TextView（默认不可见），用于隐式跳转的按钮 2（默认不可见），最下面一行是两个常规按钮，一个完成，一个取消（重置界面）。

main.xml 文件的代码如下：

```
<?xml version="1.0" encoding="utf-8"?>
<LinearLayout xmlns:android="http://schemas.android.com/apk/res/android"
    android:layout_width="fill_parent"
    android:layout_height="fill_parent"
    android:orientation="vertical"
    android:gravity="center_vertical|center_horizontal" >
<TextView
        android:layout_width="fill_parent"
```

```xml
        android:layout_height="wrap_content"
        android:text="个人信息完善"
        android:textSize="40dp"
        android:gravity="center_horizontal"/>
    <LinearLayout
        android:layout_width="fill_parent"
        android:layout_marginTop="20dp"
        android:layout_height="40dp"
        android:gravity="center_horizontal">
        <TextView
            android:layout_width="wrap_content"
            android:layout_height="fill_parent"
            android:textSize="30dp"
            android:text="姓名"
            android:gravity="center_horizontal|center_vertical"/>
        <EditText
            android:id="@+id/etName"
            android:layout_width="200dp"
            android:layout_height="fill_parent"/>
    </LinearLayout>
    <Button
        android:id="@+id/btChooseHobby"
        android:layout_width="wrap_content"
        android:layout_height="wrap_content"
        android:onClick="doClick"
        android:text="选择自己的爱好"/>
    <TextView
        android:id="@+id/tvDisplayHobby"
        android:layout_width="300dp"
        android:layout_marginTop="10dp"
        android:layout_marginBottom="10dp"
        android:textSize="30dp"
        android:layout_height="wrap_content"
```

```xml
            android:visibility="gone"/>
        <Button
            android:id="@+id/btChangeHobby"
            android:layout_width="wrap_content"
            android:layout_height="wrap_content"
            android:onClick="doClick"
            android:text="修改自己的爱好"
            android:visibility="gone"/>
        <Button
            android:id="@+id/btChooseTag"
            android:layout_width="wrap_content"
            android:layout_height="wrap_content"
            android:onClick="doClick"
            android:text="选择自己的标签"/>
        <TextView
            android:id="@+id/tvDisplayTag"
            android:layout_width="300dp"
            android:layout_height="wrap_content"
            android:layout_marginTop="10dp"
            android:textSize="30dp"
            android:layout_marginBottom="10dp"
            android:visibility="gone"/>
        <Button
            android:id="@+id/btChangeTag"
            android:layout_width="wrap_content"
            android:layout_height="wrap_content"
            android:onClick="doClick"
            android:text="修改自己的标签"
            android:visibility="gone"/>
        <LinearLayout
            android:layout_marginTop="20dp"
            android:layout_width="wrap_content"
            android:layout_height="wrap_content">
```

```xml
<Button
    android:id="@+id/btFinish"
    android:layout_width="wrap_content"
    android:layout_height="wrap_content"
    android:onClick="doClick"
    android:text="完成"/>
<Button
    android:id="@+id/btCancel"
    android:layout_marginLeft="30dp"
    android:layout_width="wrap_content"
    android:layout_height="wrap_content"
    android:onClick="doClick"
    android:text="取消"/>
    </LinearLayout>
</LinearLayout>
```

（2）在 layout 目录下新建 second.xml 文件，该文件用来描绘爱好选择界面，参见图 8-2（b）图所示，它由一个显示主题的 TextView，12 个用于选择爱好的复选按钮，1 个用于添加其他爱好的 EditText，以及 1 个提交按钮构成。

下面是显示主题的 TextView 和看小说、玩游戏、唱歌 3 个爱好复选按钮的代码段：

```xml
<?xml version="1.0" encoding="utf-8"?>
<LinearLayout xmlns:android="http://schemas.android.com/apk/res/android"
    android:layout_width="fill_parent"
    android:layout_height="fill_parent"
    android:orientation="vertical"
    android:gravity="center_vertical|center_horizontal" >
    <TextView
        android:id="@+id/tvSecondHead"
        android:layout_width="fill_parent"
        android:layout_height="wrap_content"
        android:text="李扬，您好，请选择您的爱好："
        android:textSize="20dp"
        android:layout_marginBottom="20dp"
        android:gravity="center_horizontal"/>
```

```xml
<LinearLayout
    android:layout_width="fill_parent"
    android:layout_height="wrap_content"
    android:layout_marginLeft="20dp">

    <CheckBox
        android:id="@+id/chkRead"
        android:layout_width="0dp"
        android:layout_height="wrap_content"
        android:layout_weight="1.0"
        android:text="看小说" />
    <CheckBox
        android:id="@+id/chkPlayGame"
        android:layout_width="0dp"
        android:layout_height="wrap_content"
        android:layout_weight="1.0"
        android:text="玩游戏" />
    <CheckBox
        android:id="@+id/chkSing"
        android:layout_width="0dp"
        android:layout_height="wrap_content"
        android:layout_weight="1.0"
        android:text="唱歌" />
</LinearLayout>
```

（3）参考以上代码增加踢足球、打篮球、打排球、游泳、健身、登高、静坐、聊天、思考 9 个爱好，在最下部增加其他输入框和提交按钮。

```xml
<LinearLayout
    android:layout_width="fill_parent"
    android:layout_height="wrap_content"
    android:layout_marginLeft="20dp">
    <CheckBox
        android:id="@+id/chkFootball"
        android:layout_width="0dp"
```

```xml
            android:layout_height="wrap_content"
            android:layout_weight="1.0"
            android:text="踢足球" />
        <CheckBox
            android:id="@+id/chkBasketball"
            android:layout_width="0dp"
            android:layout_height="wrap_content"
            android:layout_weight="1.0"
            android:text="打篮球" />
        <CheckBox
            android:id="@+id/chkVolleyball"
            android:layout_width="0dp"
            android:layout_height="wrap_content"
            android:layout_weight="1.0"
            android:text="打排球" />
    </LinearLayout>
    <LinearLayout
        android:layout_width="fill_parent"
        android:layout_height="wrap_content"
        android:layout_marginLeft="20dp">
        <CheckBox
            android:id="@+id/chkSwimming"
            android:layout_width="0dp"
            android:layout_height="wrap_content"
            android:layout_weight="1.0"
            android:text="游泳" />
        <CheckBox
            android:id="@+id/chkFitness"
            android:layout_width="0dp"
            android:layout_height="wrap_content"
            android:layout_weight="1.0"
            android:text="健身" />
        <CheckBox
```

```xml
            android:id="@+id/chkHeight"
            android:layout_width="0dp"
            android:layout_height="wrap_content"
            android:layout_weight="1.0"
            android:text="登高" />
</LinearLayout>
<LinearLayout
    android:layout_width="fill_parent"
    android:layout_height="wrap_content"
    android:layout_marginLeft="20dp">
    <CheckBox
            android:id="@+id/chkQuiety"
            android:layout_width="0dp"
            android:layout_height="wrap_content"
            android:layout_weight="1.0"
            android:text="静坐" />
    <CheckBox
            android:id="@+id/chkChat"
            android:layout_width="0dp"
            android:layout_height="wrap_content"
            android:layout_weight="1.0"
            android:text="聊天" />
    CheckBox
            android:id="@+id/chkPonder"
            android:layout_width="0dp"
            android:layout_height="wrap_content"
            android:layout_weight="1.0"
            android:text="思考" />
</LinearLayout>
<LinearLayout
    android:layout_width="fill_parent"
    android:layout_height="40dp"
    android:gravity="center_horizontal">
```

```xml
<TextView
    android:layout_width="wrap_content"
    android:layout_height="fill_parent"
    android:text="其他"
    android:gravity="center_horizontal|center_vertical"/>
    <EditText
        android:id="@+id/etOtherHobby"
        android:layout_width="200dp"
        android:layout_height="fill_parent"/>
</LinearLayout>
<Button
    android:id="@+id/btHobbyFinish"
    android:layout_marginTop="20dp"
    android:layout_width="wrap_content"
    android:layout_height="wrap_content"
    android:onClick="doClick"
    android:text="提交"/>
</LinearLayout>
```

（4）在 layout 目录下新建 third.xml 文件，该文件用来描述标签选择界面（参照图 8-2（c）图所示），该界面与爱好选择界面类似，可参考爱好选择界面设计该界面，该界面的标签包括 90 后、80 后、00 后、个性、哲理、唯美、搞笑、孤独、犀利、回忆、非主流、校园。代码如下：

```xml
<?xml version="1.0" encoding="utf-8"?>
<LinearLayout xmlns:android="http://schemas.android.com/apk/res/android"
    android:layout_width="fill_parent"
    android:layout_height="fill_parent"
    android:orientation="vertical"
    android:gravity="center_vertical|center_horizontal" >
<TextView
        android:id="@+id/tvThirdHead"
        android:layout_width="fill_parent"
        android:layout_height="wrap_content"
        android:textSize="20dp"
```

```xml
            android:layout_marginBottom="20dp"
            android:gravity="center_horizontal"/>
<LinearLayout
        android:layout_width="fill_parent"
        android:layout_height="wrap_content"
        android:layout_marginLeft="20dp">
        <CheckBox
            android:id="@+id/chk90s"
            android:layout_width="0dp"
            android:layout_height="wrap_content"
            android:layout_weight="1.0"
            android:text="90后" />
        <CheckBox
            android:id="@+id/chk80s"
            android:layout_width="0dp"
            android:layout_height="wrap_content"
            android:layout_weight="1.0"
            android:text="80后" />
        <CheckBox
            android:id="@+id/chk00s"
            android:layout_width="0dp"
            android:layout_height="wrap_content"
            android:layout_weight="1.0"
            android:text="00后" />
</LinearLayout>
<LinearLayout
        android:layout_width="fill_parent"
        android:layout_height="wrap_content"
        android:layout_marginLeft="20dp">
        <CheckBox
            android:id="@+id/chkPersonality"
            android:layout_width="0dp"
            android:layout_height="wrap_content"
```

```xml
            android:layout_weight="1.0"
            android:text="个性" />
        <CheckBox
            android:id="@+id/chkPhilosophy"
            android:layout_width="0dp"
            android:layout_height="wrap_content"
            android:layout_weight="1.0"
            android:text="哲理" />
        <CheckBox
            android:id="@+id/chkAestheticism"
            android:layout_width="0dp"
            android:layout_height="wrap_content"
            android:layout_weight="1.0"
            android:text="唯美" />
</LinearLayout>
<LinearLayout
        android:layout_width="fill_parent"
        android:layout_height="wrap_content"
        android:layout_marginLeft="20dp">
    <CheckBox
            android:id="@+id/chkFunny"
            android:layout_width="0dp"
            android:layout_height="wrap_content"
            android:layout_weight="1.0"
            android:text="搞笑" />
    <CheckBox
            android:id="@+id/chkloneliness"
            android:layout_width="0dp"
            android:layout_height="wrap_content"
            android:layout_weight="1.0"
            android:text="孤独" />
    <CheckBox
            android:id="@+id/chkSharp"
```

```xml
            android:layout_width="0dp"
            android:layout_height="wrap_content"
            android:layout_weight="1.0"
            android:text="犀利" />
    </LinearLayout>
    <LinearLayout
        android:layout_width="fill_parent"
        android:layout_height="wrap_content"
        android:layout_marginLeft="20dp">
        <CheckBox
            android:id="@+id/chkMemory"
            android:layout_width="0dp"
            android:layout_height="wrap_content"
            android:layout_weight="1.0"
            android:text="回忆" />
        <CheckBox
            android:id="@+id/chkAlternative"
            android:layout_width="0dp"
            android:layout_height="wrap_content"
            android:layout_weight="1.0"
            android:text="非主流" />
        <CheckBox
            android:id="@+id/chkCampus"
            android:layout_width="0dp"
            android:layout_height="wrap_content"
            android:layout_weight="1.0"
            android:text="校园" />
    </LinearLayout>
    <LinearLayout
        android:layout_width="fill_parent"
        android:layout_height="40dp"
        android:gravity="center_horizontal">
        <TextView
```

```xml
            android:layout_width="wrap_content"
            android:layout_height="fill_parent"
            android:text="其他"
            android:gravity="center_horizontal|center_vertical"/>
        <EditText
            android:id="@+id/etOtherTag"
            android:layout_width="200dp"
            android:layout_height="fill_parent"/>
    </LinearLayout>
    <Button
        android:id="@+id/btTagFinish"
        android:layout_marginTop="20dp"
        android:layout_width="wrap_content"
        android:layout_height="wrap_content"
        android:onClick="doClick"
        android:text="提交"/>
</LinearLayout>
```

3. Java 文件的开发

创建3个Activity，分别是IntentActivity、SecondActivity和ThirdActivity，这三个Activity分别加载 main.xml、second.xml 和 third.xml 这三个布局文件。

（1）IntentActivity 加载主界面，它是默认程序启动时加载的 Activity，还提供了通过 Intent 启动另外两个 Activity 的按钮。在主界面中单击【选择自己的爱好】按钮会进入爱好选择界面，这里为显式 Intent 启动。

```java
case R.id.btChangeHobby:
    //通过【选择自己的爱好】按钮进入爱好选择界面
    second_identification=2;
    second_intent = new Intent(this,SecondActivity.class);
    second_intent.putExtra("second_identification", second_identification);
    second_intent.putExtra("etName_string", etName_string);
    second_intent.putExtra("hobby_number", hobby_number);
    second_intent.putExtra("etOtherHobby_string", etOtherHobby_string);
    startActivityForResult(second_intent, SECOND_REQUESTCODE);
```

```
        break;
```

通过【选择自己的标签】按钮进入标签选择界面，这里为隐式 Intent 启动。

```
case R.id.btChooseTag:
    //通过【选择自己的标签】按钮进入标签选择界面
    third_identification=1;
    third_intent.setAction("com.xunfang.action");
    third_intent.addCategory("com.xunfang.category");
    third_intent.putExtra("third_identification", third_identification);
    third_intent.putExtra("etName_string", etName_string);
    startActivityForResult(third_intent, THIRD_REQUESTCODE);
    break;
```

获取到 Activity 返回值后，进行相应的操作。

```
//获取到 Activity 的返回值后的处理方法
@Override
protected void onActivityResult(int requestCode, int resultCode, Intent data) {
    // TODO Auto-generated method stub
    switch(requestCode){
    case SECOND_REQUESTCODE:
        //隐藏选择爱好按钮，显示爱好界面和修改爱好按钮
        btChooseHobby.setVisibility(View.GONE);
        tvDisplayHobby.setVisibility(View.VISIBLE);
        btChangeHobby.setVisibility(View.VISIBLE);
        tvDisplayHobby.setText(data.getStringExtra("hobby_string"));
        hobby_number=data.getStringExtra("hobby_number");
        etOtherHobby_string=data.getStringExtra("etOtherHobby_string");
        break;
    case THIRD_REQUESTCODE:
        btChooseTag.setVisibility(View.GONE);
        tvDisplayTag.setVisibility(View.VISIBLE);
        btChangeTag.setVisibility(View.VISIBLE);
        tvDisplayTag.setText(data.getStringExtra("tag_string"));
        tag_number=data.getStringExtra("tag_number");
        etOtherTag_string=data.getStringExtra("etOtherTag_string");
```

```
        break;
    }
    super.onActivityResult(requestCode, resultCode, data);
}
```

（2）SecondActivity.Java 文件对应爱好选择界面，该 Activity 将通过显式启动的方式启动，且将会在退出后将数据返回原 Activity。

```
//设置返回值
Intent result = new Intent();
result.putExtra("hobby_number",hobby_number);
result.putExtra("hobby_string", hobby_string);
result.putExtra("etOtherHobby_string", etOtherHobby_string);
setResult(RESULT_OK, result);
finish();
break;
```

（3）ThirdActivity.Java 文件以 third.xml 文件为界面。该 Activity 通过隐式启动的方式启动，同样，该 Activity 退出后也要将数据返回原 Activity。

```
//设置返回值
Intent result = new Intent();
result.putExtra("tag_number",tag_number);
result.putExtra("tag_string", tag_string);
result.putExtra("etOtherTag_string", etOtherTag_string);
setResult(RESULT_OK, result);
finish();
break;
```

程序编写好后，将其运行到 Android 开发终端上进行测试。

四、项目思考与扩展

1. 应用 Intent 的传值，开发如图 8-4 所示的身高体重计算器。

图 8-4　身高体重计算器效果图

2. 使用隐性 Intent，进行网页的打开。

项目九　Activity 的生命周期

【本章导读】

Android 系统中包含很多的组件，其中有 4 个组件在 Android 系统中最为重要，分别是 Activity、Service、BroadcastReceiver 和 ContentProvider。Activity 是 Android 程序的呈现层，显示可视化的用户界面，并接收与用户交互所产生的界面事件，更具体的来说我们在应用程序中看到的每个窗口就是一个 Activity。Service 一般用于没有用户界面，但需要长时间在后台运行的应用，一般都是后台中需要运行的程序。BroadcastReceiver 是用来接收并响应广播消息的组件。ContentProvider 是 Android 系统提供的一种标准的数据共享机制。

Activity 是 Android 最为重要的组件之一，中文的意思是活动，在 Android 程序中，Activity 代表手机屏幕的一屏，或是平板电脑中的一个窗口。每一个窗口从创建到销毁都有代表其具体的生命周期的一个活动，本章的学习项目就是来学习一下 Activity 的生命周期的运行规律。

一、项目要求

1. 掌握 Activity 生命周期函数的用法。
2. 掌握 Log 的使用。
3. 重写 Activity 生命周期的事件回调函数，在其中加 Log 输出。
4. 运行 Android 工程，在 Logcat 中查看 Activity 生命周期事件回调函数的执行步骤。

二、项目相关知识

1. Android 生命周期与进程优先级

Android 生命周期是从程序启动到程序终止的全过程。Android 系统中的进程有优先级的区别，用于有序化系统对内存的回收。进程从高优先级到低优先级依次为前台进程、可见进程、服务进程、后台进程和空进程。优先级低的，在系统资源不足时，先被回收。

- **前台进程**：前台进程是 Android 系统中最重要的进程，是与用户正在交互的进程。
- **可见进程**：可见进程指部分程序界面能够被用户看见，却不在前台与用户交互，不响应界面事件的进程。
- **服务进程**：服务进程是包含启动服务的进程。
- **后台进程**：如果一个进程不包含任何已经启动的服务，而且没有任何用户可见的 Activity，则这个进程就是后台进程。
- **空进程**：指不包含任何活跃组件的进程。

2. Activity 的生命周期

Activity 的生命周期指 Activity 从启动到销毁的过程，在这个过程中，Activity 一般表现为 4 种状态，分别是活动状态、暂停状态、停止状态和非活动状态。其执行的函数方法和生命周期流程如图 9-1 所示。

下面具体解释一下 Android 生命周期的几个过程：

（1）启动 Activity：系统会先调用 onCreate 方法，然后调用 onStart 方法，最后调用 onResume，Activity 进入运行状态。

（2）当前 Activity 被其他 Activity 覆盖其上或被锁屏：系统会调用 onPause 方法，暂停当前 Activity 的执行。

（3）当前 Activity 由被覆盖状态回到前台或解锁屏：系统会调用 onResume 方法，再次进入运行状态。

（4）当前 Activity 转到新的 Activity 界面或按 Home 键回到主屏，自身退居后台：系统会先调用 onPause 方法，然后调用 onStop 方法，进入停滞状态。

项目九　Activity 的生命周期

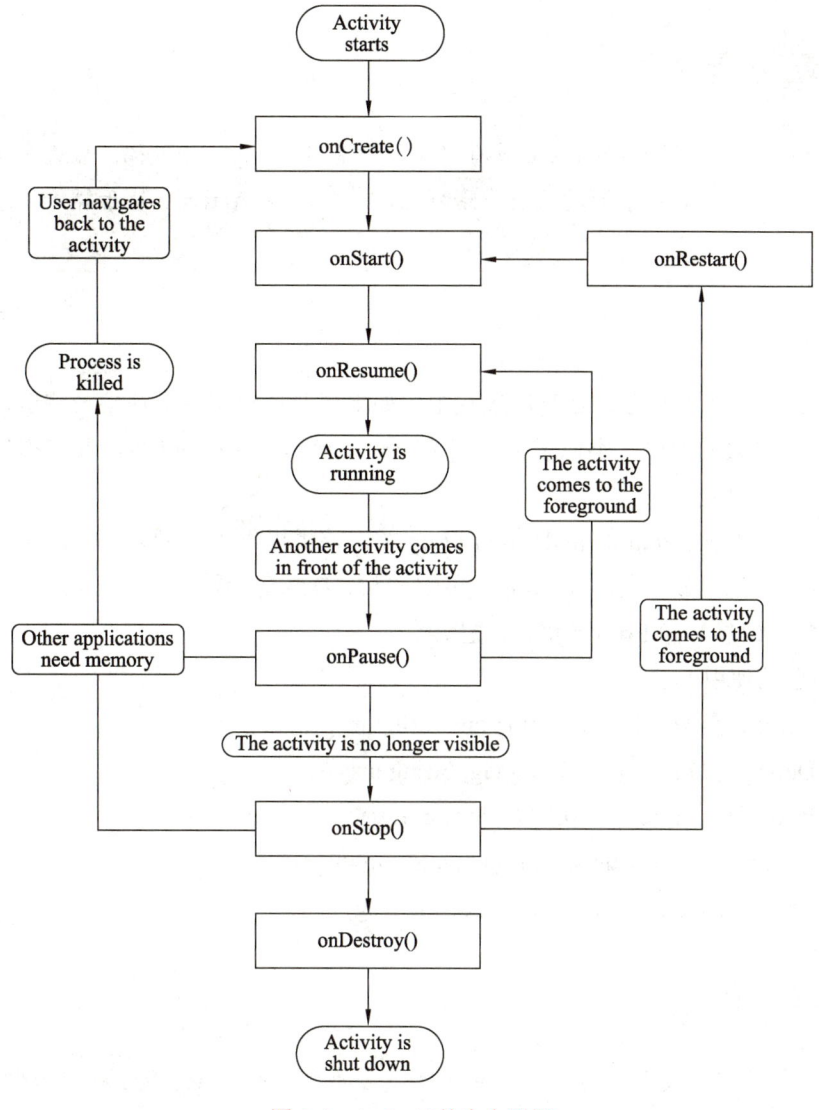

图 9-1　Android 的生命周期

（5）用户后退回到此 Activity：系统会先调用 onRestart 方法，然后调用 onStart 方法最后调用 onResume 方法，再次进入运行状态。

（6）当前 Activity 处于被覆盖状态或者后台不可见状态，即第 2 步和第 4 步，系统内存不足，杀死当前 Activity，而后用户退回当前 Activity：再次调用 onCreate 方法、onStart 方法、onResume 方法，进入运行状态。

（7）用户退出当前 Activity：系统先调用 onPause 方法，然后调用 onStop 方法，最后调用 onDestory 方法，结束当前 Activity。

> Activity 栈保存了已经启动且没有终止的所有 Activity，并遵循"后进先出"的规则。栈顶的 Activity 处于活动状态，除栈顶以外的其他 Activity 处于暂停状态或停止状态。

3. Log 类的使用

一个 Android 应用程序在运行的过程中，如果想要在控制台上观察一些运行数据，就需要打印输出结果的方法，类似于 Java 中的 print() 函数，Android 也给我们提供了类似的方法，这个方法就是 Log 类。

Log 类来自于 Android 的 android.util.Log 类，该类提供了 5 种静态方法供我们在控制台输出不同的信息，分别是：Verbose（啰嗦信息），Debug（调试信息），Info（一般信息），Warning（警告信息）和 Error（错误信息）。

5 种信息的使用方法如下：

（1）Verbose 信息：Log.v (String tag, String msg)；

（2）Debug 信息：Log.d (String tag, String msg)；

（3）Info 信息：Log.i (String tag, String msg)；

（4）Warning 信息：Log.w (String tag, String msg)；

（5）Error 信息：Log.e (String tag, String msg)。

> tag 是一个标识变量，可以是任意一个定义的字符串，msg 表示输出的信息内容。

三、项目实施过程

下面我们通过一个例子来体会 Activity 的生命周期过程。

（1）创建名为 AndroidCode09 的 Android 工程，包结构为"com.xdxy.cycle"，Activity 名为 FirstActivity，如图 9-2 所示。

项目九　Activity 的生命周期

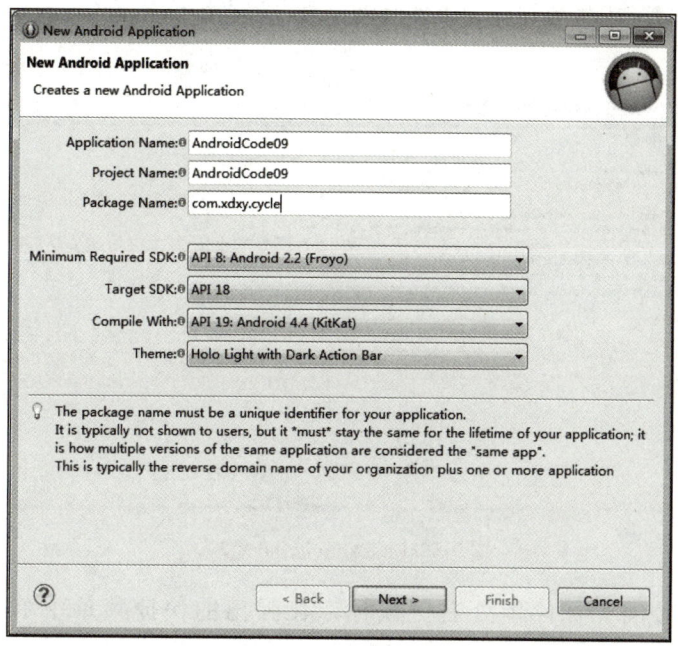

图 9-2　创建工程

（2）如图 9-3 所示，在 src 目录下找到 FirstActivity.Java 文件，双击打开，如图 9-4 所示。

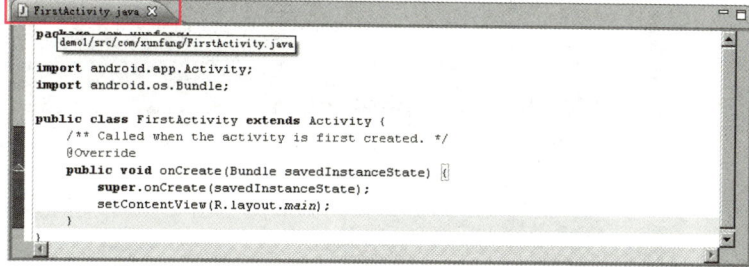

图 9-3　工程目录　　　　　　　　　　图 9-4　FirstActivity.Java 文件

用鼠标左键双击图 9-4 中红框处可将编辑框放到最大，再次双击可还原。

（3）添加 Activity 生命周期的事件回调函数，将光标停到图 9-5 红框内所示处后按回车键。

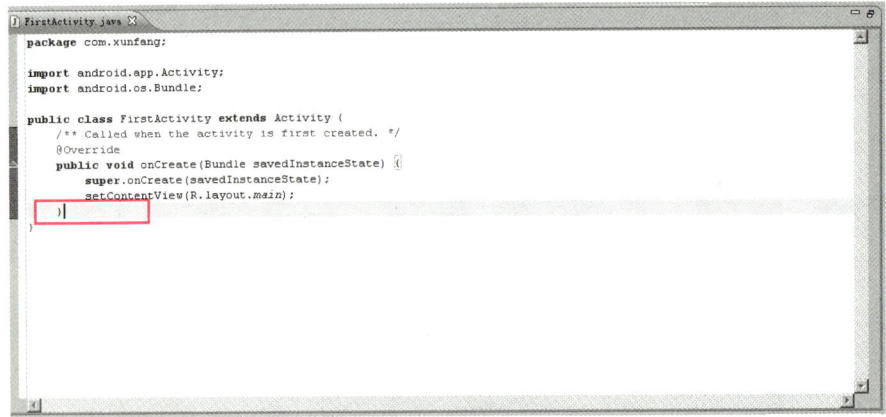

图 9-5　FirstActivity.Java 文件

（4）在此时的光标停留处右键鼠标，在弹出的快捷菜单中单击【Source】→【Override/ImplementMethods...】项，如图 9-6 所示。

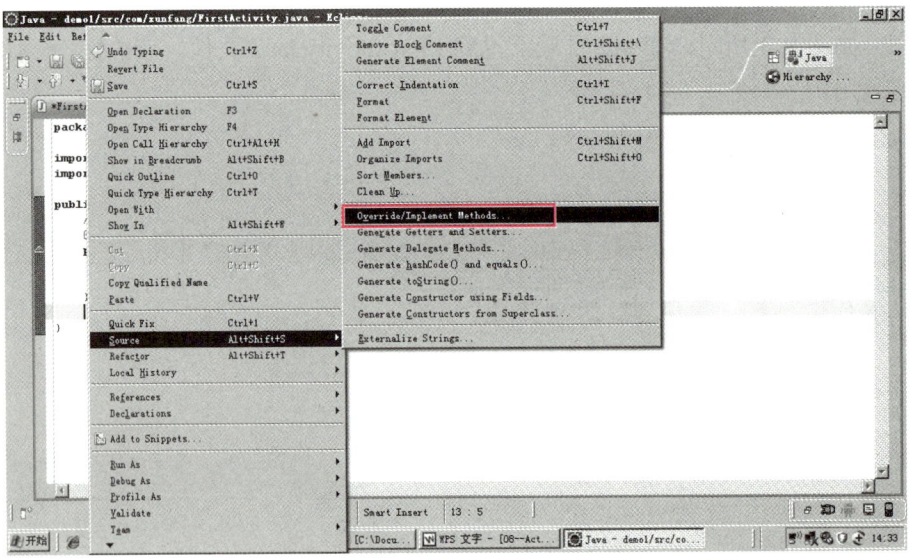

图 9-6　单击图示项

（5）在弹出的 Override/ImplementMethods 对话框中，拖动右侧滑动条，找到 onDestroy()项并选中，如图 9-7 所示。

项目九　Activity 的生命周期

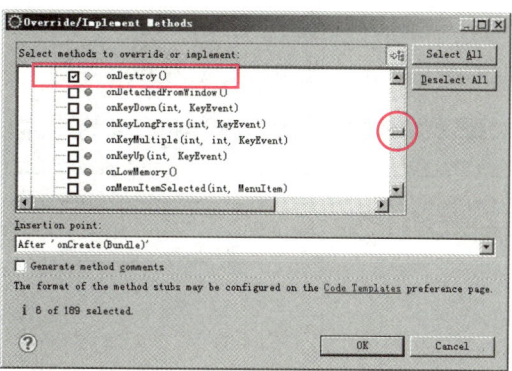

图 9-7　选中 onDestroy()项

（6）按照同样方法在图 9-7 中选中 onPause()、onRestart()、onResume()、onStart()和 onStop()，单击【OK】按钮，此时代码中将会出现我们刚刚选择的函数，如图 9-8 所示。

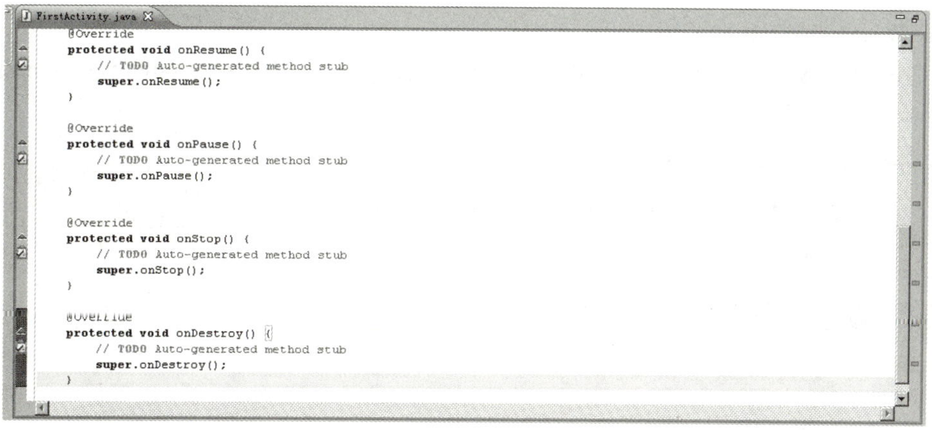

图 9-8　FirstActivity.Java 文件

（7）在 FirstActivity 类中添加代码，设置一个用于查看结果的字符串标记。

public class FirstActivity extends Activity {
　　/** Called when the activity is first created. */
　　String TAG="FirstActivity";
}

（8）在事件回调函数中添加 Log 输出，在 onCreate 方法中添加代码。

public void onCreate(Bundle savedInstanceState) {
　　　　super.onCreate(savedInstanceState);
　　　　Log.i("info",TAG+".onCreate()");
　　}

（9）如果 Log 下面有一条红线，则按【Ctrl+Shift+O】快捷键进入 OrganizeImports 对话框，双击图 9-9 红框中条目退出，此步目的是为 Log 导包。

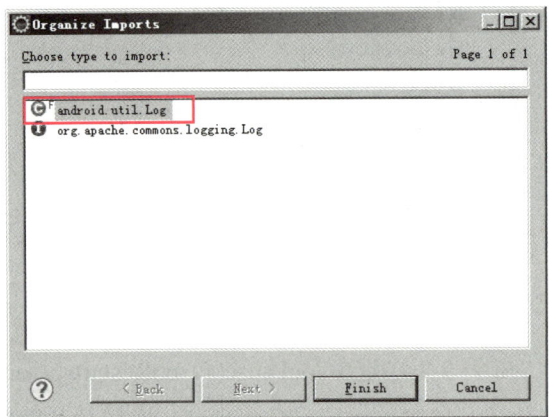

图 9-9　导包

（10）与（8）同样的方法为 onStart()、onRestart()、onResume()、onPause()、onStop()、onDestroy()方法添加 Log 输出代码。

```
    protected void onStart() {
        // TODO Auto-generated method stub
        super.onStart();
        Log.i("info",TAG+".onStart()");
    }
    @Override
    protected void onRestart() {
        // TODO Auto-generated method stub
        super.onRestart();
        Log.i("info",TAG+".onRestart()");
    }
    @Override
    protected void onResume() {
        // TODO Auto-generated method stub
        super.onResume();
        Log.i("info",TAG+".onResume()");
    }
    @Override
```

项目九　Activity 的生命周期

```
protected void onPause() {
    // TODO Auto-generated method stub
    super.onPause();
    Log.i("info",TAG+".onPause()");
}

@Override
protected void onStop() {
    // TODO Auto-generated method stub
    super.onStop();
    Log.i("info",TAG+".onStop()");
}
@Override
protected void onDestroy() {
    // TODO Auto-generated method stub
    super.onDestroy();
    Log.i("info",TAG+".onDestroy()");
}
```

（11）下面创建新类 SecondActivity，选中包后，右击鼠标，在弹出的快捷菜单中选择【New】→【Class】项，如图 9-10 所示。

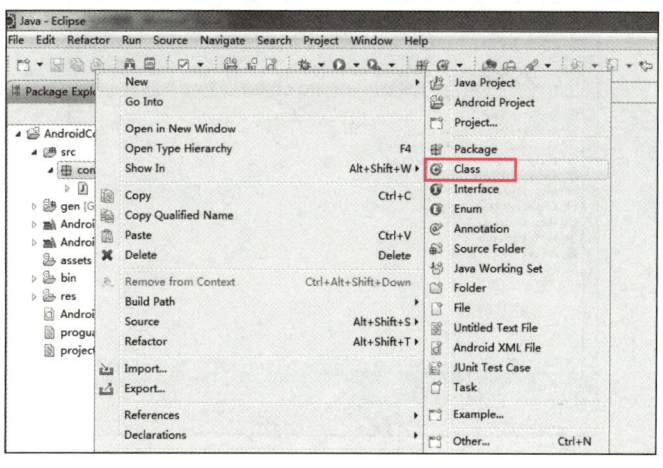

图 9-10　单击 Class 项

（12）如图 9-11 所示，在 New Java Class 对话框中填入类名 SecondActivity，单击

123

【Browse...】按钮选择父类。

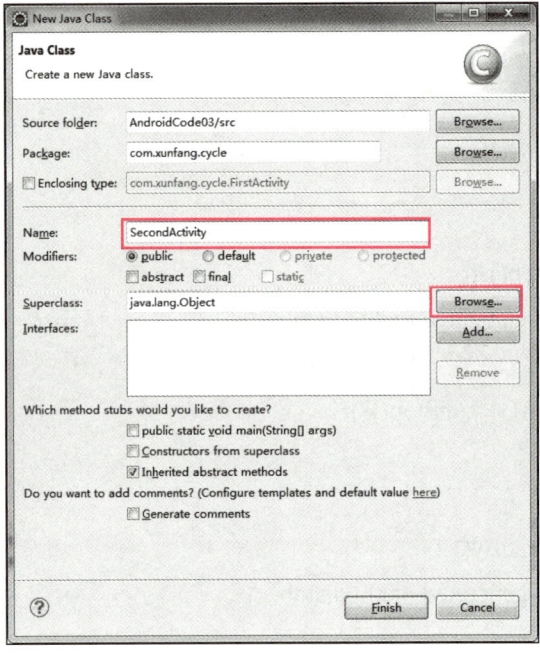

图 9-11　New Java Class 对话框

（13）弹出的选择父类对话框如图 9-12 所示，在搜索栏中输入 activity，此时下面将会出现 android.app 包下的一个 Activity 类，选中后，单击【OK】按钮。

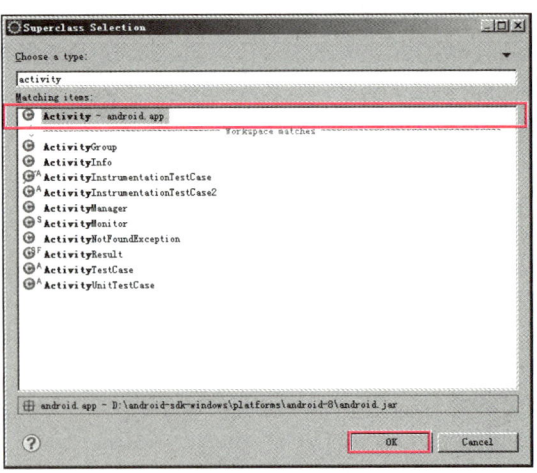

图 9-12　选择父类对话框

（14）此时进入如图 9-13 所示画面，单击【Finish】完成创建。

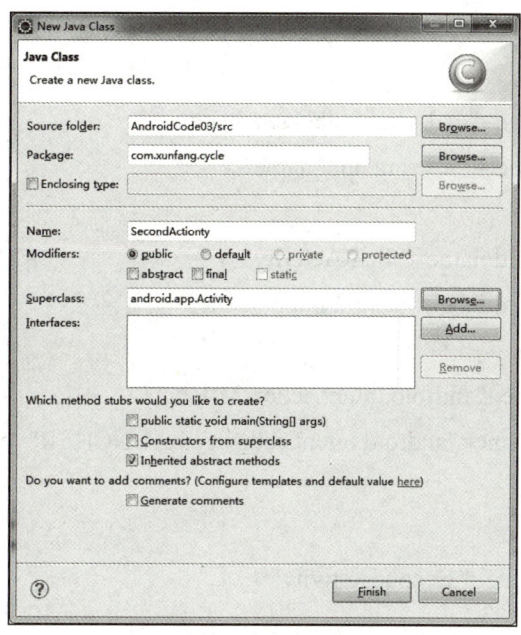

图 9-13　单击 Finish

（15）将 FirstActivity 中的代码拷贝到 SecondActivity 中，修改代码，然后按【Ctrl+S】组合键保存。

```
public class SecondActivity extends Activity {
/** Called when the activity is first created. */
String TAG="SecondActivity";
    }
```

（16）打开 AndroidManifest.xml 文件，为 SecondActivity 注册。首先如图 9-14 所示双击 AndroidManifest.xml 打开该文件，然后切换至代码显示页面，如图 9-15 所示。

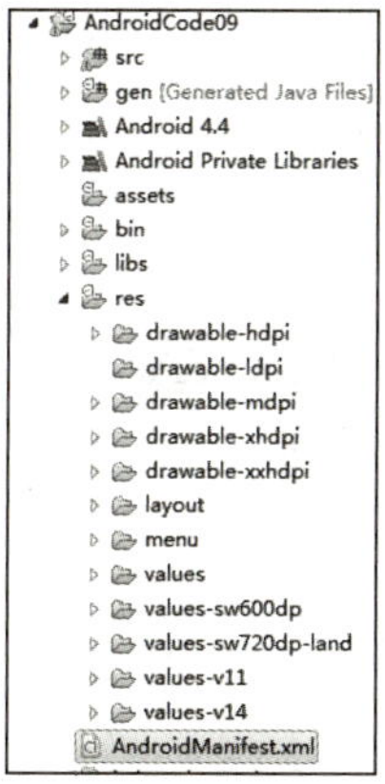

图 9-14　打开文件 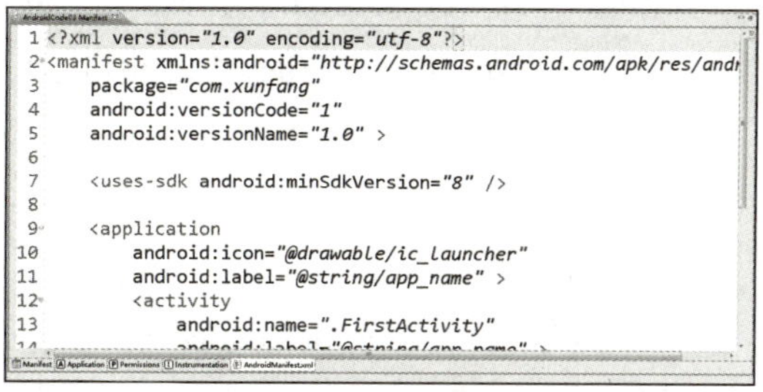　　　　　　　　图 9-15　打开代码编辑界面

（17）添加代码，为 SecondActivity 注册，然后按【Ctrl+S】组合键保存。

```xml
<application
        android:icon="@drawable/ic_launcher"
        android:label="@string/app_name" >
<activity
        android:name=".FirstActivity"
        android:label="@string/app_name" >
<intent-filter>
<action android:name="android.intent.action.MAIN" />
<category android:name="android.intent.category.LAUNCHER" />
</intent-filter>
</activity>
<activity android:name=".SecondActivity"/>
</application>
```

（18）在 FirstActivity 的 onCreate 方法中输入代码。

```java
public class FirstActivity extends Activity {
    /** Called when the activity is first created. */
    String TAG="FirstActivity";
    @Override
    public void onCreate(Bundle savedInstanceState) {
        super.onCreate(savedInstanceState);
        Button button = new Button(this);
        button.setText("单击按钮");
        button.setOnClickListener(new OnClickListener() {
            @Override
            public void onClick(View v) {
                // TODO Auto-generated method stub
                startActivity(new Intent(FirstActivity.this,SecondActivity.class));
            }
        });
        setContentView(button);
        Log.i("info",TAG+".onCreate()");
    }
```

(19)按【Ctrl+Shift+O】组合键导包,单击图 9-16 红框中条目,单击【Finish】按钮,然后保存。

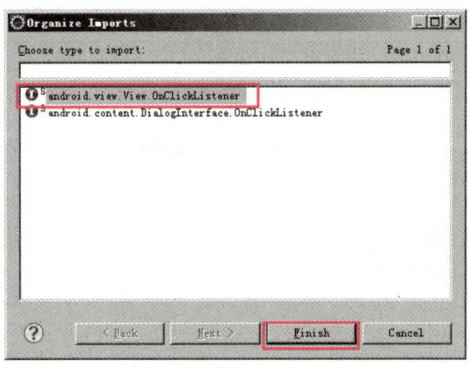

图 9-16 导包

(20)参照第(18)步操作修改 SecondActivity。

(21)至此工程编写完成,打开 LogCat,单击图 9-17 中红框内的绿色加号添加过滤器。

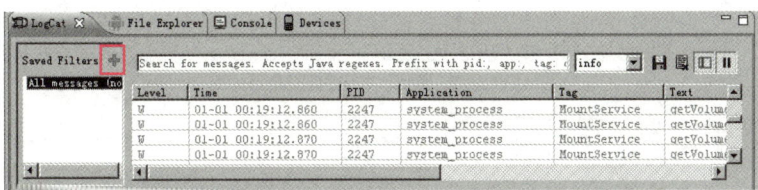

图 9-17 单击红框内的绿色加号

(22)在弹出的对话框中输入或选定如图 9-18 所示项目,单击【OK】按钮。

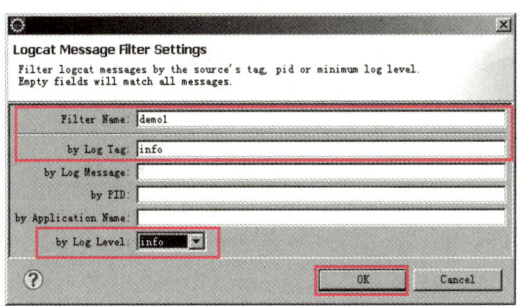

图 9-18 单击 OK

(23)将工程运行到 Android 开发终端上,此时 LogCat 的输出如图 9-19 所示。

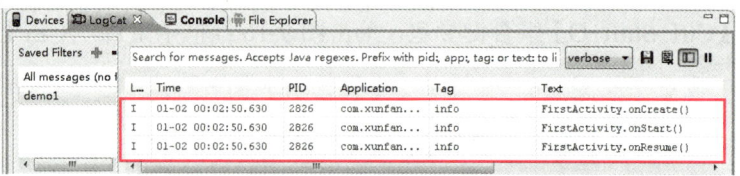

图 9-19　LogCat 输出信息

（24）打开 SecondActivity 后，LogCat 输出如图 9-20 红框内所示内容。

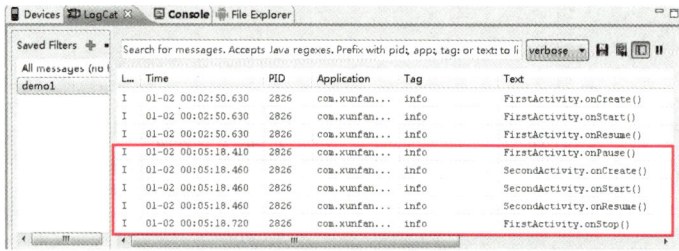

图 9-20　Log 输出

（25）关闭 SecondActivity，此时 LogCat 输出如图 9-21 红框内所示内容。

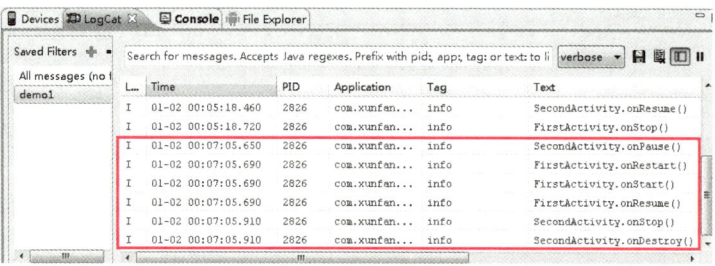

图 9-21　Log 输出

（26）按 Android 开发终端上的返回键退出程序，此时 Log 输出如图 9-22 红框中所示内容。

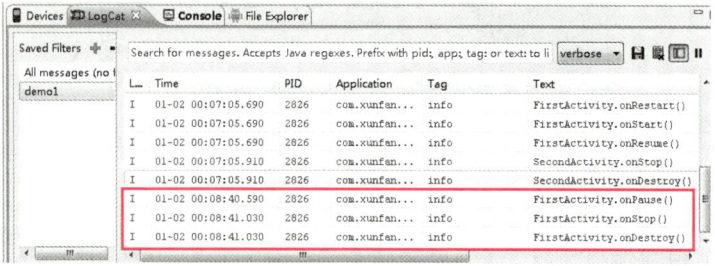

图 9-22　Log 输出

四、项目思考与扩展

1. 手动创建一个 Activity 作为程序启动的第一个 Activity，如图 9-23 所示。

图 9-23　新建 activity

2. 在程序中加入类似 log.d("debug"，"1")的代码，在 LogCat 中打印输出消息，按照 Activity 中各回调函数执行的顺序，依次在 LogCat 中打印输出 1、2、3、4、5、6，如图 9-24 所示。

Level	Time	PID	TID	Application	Tag	Text
D	07-01 15:26:02.200	29342	29342		debug	1
D	07-01 15:26:02.200	29342	29342		debug	2
D	07-01 15:26:02.200	29342	29342		debug	3
D	07-01 15:27:44.450	29342	29342	com.xunfang.cycle	debug	4
D	07-01 15:27:44.590	29342	29342	com.xunfang.cycle	debug	5
D	07-01 15:27:44.590	29342	29342	com.xunfang.cycle	debug	6

图 9-24　Activity 执行顺序

进阶篇

项目十 高级用户界面设计

【本章导读】

在前几个项目中，我们已经掌握 Android 应用程序中基本组件的使用方法，并学习了 Android 中的 Activity 类的编写。在一些项目中为了实现一些特殊效果，需要一些高级组件来丰富 UI 的效果，本章项目我们就来学习自动完成文本框、消息提示框、对话框等一些高级组件的设计和开发。

一、项目要求

1. 掌握自动完成文本框 AutoCompleteTextView 的使用方法。
2. 掌握进度条 ProgressBar 的使用方法。
3. 掌握拖动条 SeekBar 的使用方法。
4. 掌握星级评分条 RatingBar 的使用方法。
5. 掌握选项卡 TabHost 的使用方法。
6. 掌握图像切换器 ImageSwitcher 的使用方法。
7. 掌握画廊视图 Gallery 的使用方法。
8. 掌握消息提示框 Toast 的使用方法。
9. 掌握实训项目仿 Windows 7 图片预览窗格效果的实现方法。

二、项目相关知识

1. 自动完成文本框 AutoCompleteTextView

自动完成文本框（AutoCompleteTextView），可根据用户输入的内容，匹配指定的数据源，以列表的形式显示数据源中所有符合要求的数据，以供用户选择，减少用户的输入内容，方便用户使用。例如常用的百度搜索。

自动完成文本框的定义格式如下：

```
<AutoCompleteTextView
属性列表
>
</AutoCompleteTextView>
```

在 Android 中 AutoCompleteTextView 组件类继承自 EditText 类，所以 EditText 组件所支持的属性，AutoCompleteTextView 组件类都可以进行直接的调用。AutoCompleteTextView 支持的 XML 属性如表 10-1 所示。

表 10-1 AutoCompleteTextView 支持的 XML 属性

XML 属性	描述
android:completionHint	为弹出的下拉菜单指定提示标题
android:completionThreshold	指定用户至少输入几个字符才会显示提示，默认为 2
android:dropDownHeight	指定下拉菜单的高度
android:dropDownHorizontalOffset	指定下拉菜单与文本之间的水平偏移，下拉菜单默认与文本框左对齐
android:dropDownVerticalOffset	指定下拉菜单与文本之间的垂直偏移，下拉菜单默认紧跟文本框
android:dropDownWidth	指定下拉菜单的宽度
android:popupBackGround	为下拉菜单设置背景
android:ems	设置输入字符的长度。当设置该属性后，控件显示的长度就为此长度，超出的部分将不显示

高级组件的程序开发与一般组件程序开发有所不同。在高级组件开发中，一般需要 XML 文件和 Java 程序的配合使用才能完成，下面是一个带自动提示功能的搜索框的实现，程序如下所示：

1）布局文件 activity_main.xml

程序代码如下：

```xml
<LinearLayout xmlns:android="http://schemas.android.com/apk/res/android"
    android:layout_width="match_parent"
    android:layout_height="match_parent"
    android:orientation="horizontal" >
    <AutoCompleteTextView
        android:id="@+id/autoCompleteTextView1"
        android:layout_width="wrap_content"
        android:layout_height="wrap_content"
        android:layout_weight="7"
        android:completionHint="输入搜索内容"     //在下拉菜单指定提示标题
        android:completionThreshold="1"         //指定用户至少输入几个字符才会显示提示
        android:text="" >
    </AutoCompleteTextView>
    <Button
        android:id="@+id/button1"
        android:layout_width="wrap_content"
        android:layout_height="wrap_content"
        android:layout_marginLeft="10dp"
        android:layout_weight="1"
        android:text="搜索" />
</LinearLayout>
```

2）主活动文件 MainActivity.Java 程序

```java
public class MainActivity extends Activity {
    /** 成员变量：字符串数组，用于保存下拉菜单列表项 */
    private static final String[] COUNTRIES = { "天猫","天气预报","天天向上","天津",
                                                "天津小吃","天津现代职业技术学院",
                                                "美丽天津" };
    @Override
    protected void onCreate(Bundle savedInstanceState) {
        super.onCreate(savedInstanceState);
        setContentView(R.layout.activity_main);
        // 获取自动完成文本框
        final AutoCompleteTextView autoCompleteTextView
```

```
        = (AutoCompleteTextView) findViewById(R.id.autoCompleteTextView1);
// 创建一个 ArrayAdapter 适配器，显示列表项
ArrayAdapter<String> adapter = new ArrayAdapter<String>(this,
                android.R.layout.simple_dropdown_item_1line, COUNTRIES);
// 自动完成文本框与适配器关联
autoCompleteTextView.setAdapter(adapter);
// 获取按钮，并添加单击事件监听器
Button button = (Button) findViewById(R.id.button1);
button.setOnClickListener(new OnClickListener() {
@Override
    public void onClick(View arg0) {
// 通过消息提示框，显示自动完成文本框中输入的内容
        Toast.makeText(MainActivity.this,
        autoCompleteTextView.getText().toString(),
        Toast.LENGTH_SHORT).show();
    }
});
 }
}
```

实现的效果如图 10-1 所示。

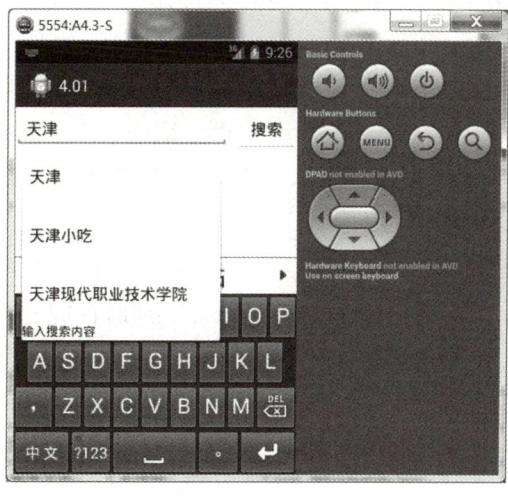

图 10-1　AutoCompleteTextView 实现效果图

2. 进度条 ProgressBar

ProgressBar 也是 UI 界面中常用的一种高级组件，通常用于向用户显示某些耗时操作完成的百分比。在实际应用中，当一个应用程序在后台执行项目时（如下载项目），前台界面不会有任何信息，这时用户根本不知道程序是否在执行，以及执行的进度等，因此需要使用 ProgressBar 来显示程序执行的进度。

> 另外，还有一种进度条，可以直接在窗口标题上显示，这种进度条不需要使用 ProgressBar 组件，它可以直接由 Activity 方法进行启动和使用。

在开发时，我们可以通过 style 属性为 ProgressBar 指定风格，该属性支持如下几种属性值：

- @android:style/Widget.ProgressBar.Horizontal：水平进度条。
- @android:style/Widget.ProgressBar.Inverse：细的、顺时针旋转的、中等圆形普通进度条。
- @android:style/Widget.ProgressBar.Large：粗的、顺时针旋转的、大圆形进度条。
- @android:style/Widget.ProgressBar.Large.Inverse：粗的、顺时针旋转的、大圆形普通进度条。
- @android:style/Widget.ProgressBar.Small：细的、顺时针旋转的、小圆形进度条。
- @android:style/Widget.ProgressBar.Small.Inverse：细的、顺时针旋转的、小圆形普通进度条。
- "?android:attr/progressBarStyle"：逆时针旋转的、半封闭的、中等圆形进度条。
- "?android:attr/progressBarStyleHorizontal"：细的、水平方向的进度条。
- "?android:attr/progressBarStyleInverse"：逆时针旋转的、半封闭的、中等圆形普通进度条。
- "?android:attr/progressBarStyleLarge"：逆时针旋转的、全封闭的、大圆形进度条。
- "?android:attr/progressBarStyleLargeInverse"：逆时针旋转的、全封闭的、大圆形普通进度条。
- "?android:attr/progressBarStyleSmall"：逆时针旋转的、半封闭的、小圆形进度条。
- "?android:attr/progressBarStyleSmallInverse"：逆时针旋转的、半封闭的、小圆形进度条。

ProgressBar 在开发的时候支持如表 10-2 所示的属性。

表 10-2　ProgressBar 支持的 XML 属性

XML 属性	说　明
android:max	设置该进度条的最大值
android:progress	设置该进度条的已完成进度值
android: progressDrawable	设置该进度条的轨道的绘制形式
android:mindeterminate	若该属性设为 true，则进度条不精确显示进度
android:mindeterminateDrawable	设置绘制不显示进度的进度条的 Drawable 对象
android:mindeterminateDuration	该属性不精确显示进度的持续时间

ProgressBar 的常用方法有以下两个：

➢ SetProgress(int)：设置进度的完成百分比。

➢ IncrementProgressBy(int)：设置进度条的进度增加或减少量。参数为整数增，为负数减。

　　在进行开发的时候，如果需要在标题上显示的进度条，我们可以首先调用 Activity 的 requestWindowFeature()方法，然后传入参数启用特定的窗口特征，其参数选择分为带进度的进度 Window.FEATURE_PROGRESS 和不带进度的进度 Window.FEATURE_INDETERMINATE_PROGRESS 的两种参数，然后调用 ActivitysetProgressBarVisibility(boolean)方法即可完成操作。

实现一个水平进度条和圆形进度条的程序如下：

1）布局文件 activity_main.xml

```
<LinearLayout xmlns:android="http://schemas.android.com/apk/res/android"
    android:layout_width="match_parent"
    android:layout_height="match_parent"
    android:orientation="vertical" >
    <!-- 水平进度条 -->
    <ProgressBar
        android:id="@+id/progressBar1"
        style="@android:style/Widget.ProgressBar.Horizontal"     //水平进度条
        android:layout_width="match_parent"
        android:layout_height="wrap_content"
```

```xml
            android:max="100" />
        <!-- 圆形进度条 -->
        <ProgressBar
            android:id="@+id/progressBar2"
            style="?android:attr/progressBarStyleLarge"        //大圆形进度条
            android:layout_width="wrap_content"
            android:layout_height="wrap_content" />
</LinearLayout>
```

2）主活动文件 MainActivity.Java

```java
public class MainActivity extends Activity {
    /** 成员变量：4个 */
    private ProgressBar horizonP;                //水平进度条
    private ProgressBar circleP;                 //圆形进度条
    private int mProgressStatus = 0;             //完成进度状态
    private Handler mHandler;                    //用于处理消息的 Handler 类的对象
    protected void onCreate(Bundle savedInstanceState) {
        super.onCreate(savedInstanceState);
        setContentView(R.layout.activity_main);
        // 获取水平进度条和圆形进度条
        horizonP = (ProgressBar) findViewById(R.id.progressBar1);
        circleP = (ProgressBar) findViewById(R.id.progressBar2);
        // 通过匿名内部类实例化处理消息的 Handler 类的对象，
        //并重写 handleMessage()方法，实现耗时操作没有完成时更显进度条
        mHandler = new Handler() {
            @Override
            public void handleMessage(Message msg) {
                if (msg.what == 0x111) {
                    horizonP.setProgress(mProgressStatus);        //更新进度
                } else {
                    Toast.makeText(MainActivity.this, "耗时操作已完成",
                        Toast.LENGTH_SHORT).show();
                    horizonP.setVisibility(View.GONE);
                    //设置进度条不显示，并且不占用空间
```

```
                        circleP.setVisibility(View.GONE);
                                    //设置进度条不显示,并且不占用空间
                        }
                }
        }
}
//开启一个线程,用于模拟一个耗时操作,在线程中,调用setMessage()方法发送处理消息
    new Thread(new Runnable() {
        @Override
                public void run() {
                        while (true) {
                                mProgressStatus = doWork();        //获取耗时操作完成的百分比
                                Message m = new Message();
                                if (mProgressStatus < 100) {
                                        m.what = 0x111;
                                        mHandler.sendMessage(m);        //发送消息
                                } else {
                                        m.what = 0x110;
                                        mHandler.sendMessage(m);        //发送消息
                                        break;
                                }
                        }
                }
//模拟一个耗时操作
    private int doWork() {
                mProgressStatus += Math.random() * 10;        //改变完成进度
                try {
                        Thread.sleep(200);                //线程休眠200毫秒
                } catch (InterruptedException e) {
                        e.printStackTrace();
                }
                return mProgressStatus;                //返回新的进度
        }
        }).start();                                //开启线程
```

        }
    }
实现效果如图 10-2 所示。

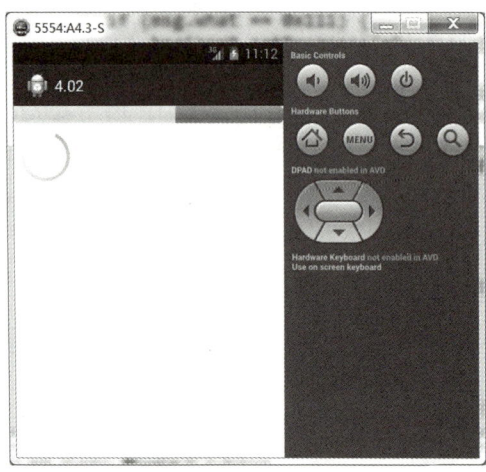

图 10-2   ProgressBar 实现效果图

3．拖动条 SeekBar

拖动条通常用于对 Android 系统中的某项数值进行调节，如音量、亮度等。拖动条和前面介绍的进度条非常相似，进度条 ProgressBar 采用颜色填充来表明进度完成程度，而拖动条 SeekBar 通过滑块的位置来标识数值，并且允许用户拖动滑块来改变进度。在开发中，程序员可通过设置 android:thumb 属性来改变滑块外观（需要制定一个 Drawable 对象，该对象作为自定义滑块）。为了让程序能够响应拖动条滑块位置的改变，我们可以为它绑定一个名为 OnSeekBarChangeListener 监听器。

下面我们来看一个在屏幕上显示拖动条的例子。

1）布局文件 activity_main.xml

```
<LinearLayout xmlns:android="http://schemas.android.com/apk/res/android"
    android:layout_width="match_parent"
    android:layout_height="match_parent"
    android:orientation="vertical" >
    <TextView
        android:id="@+id/textView1"
        android:layout_width="wrap_content"
        android:layout_height="wrap_content"
```

```xml
        android:text="当前值：50" />
    <!-- 拖动条 -->
    <SeekBar
        android:id="@+id/seekBar1"
        android:layout_width="match_parent"
        android:layout_height="wrap_content"
        android:max="100"                       //拖动条最大值
        android:padding="10px"
        android:progress="50" />                //拖动条当前值
</LinearLayout>
```

2）主活动文件 MainActivity.Java

```java
public class MainActivity extends Activity {
/** 成员变量 */
    private SeekBar seekBar;                    //拖动条
    @Override
    protected void onCreate(Bundle savedInstanceState) {
        super.onCreate(savedInstanceState);
        setContentView(R.layout.activity_main);
        // 获取文本视图
        final TextView result = (TextView) findViewById(R.id.textView1);
        // 获取拖动条，并添加拖动事件监听器
        seekBar = (SeekBar) findViewById(R.id.seekBar1);
        seekBar.setOnSeekBarChangeListener(new OnSeekBarChangeListener() {
            @Override
            public void onProgressChanged(SeekBar arg0, int arg1, boolean arg2) {
                result.setText("当前值：" + arg1);
            }
            @Override
            public void onStartTrackingTouch(SeekBar arg0) {
                Toast.makeText(MainActivity.this, "开始滑动", Toast.LENGTH_SHORT) .show();
            }
            @Override
            public void onStopTrackingTouch(SeekBar arg0) {
```

```
                Toast.makeText(MainActivity.this, "结束滑动"+
                           arg0.getProgress(), Toast.LENGTH_SHORT) .show();
            }
        });
    }
}
```

SeekBar 的效果如图 10-3 所示。

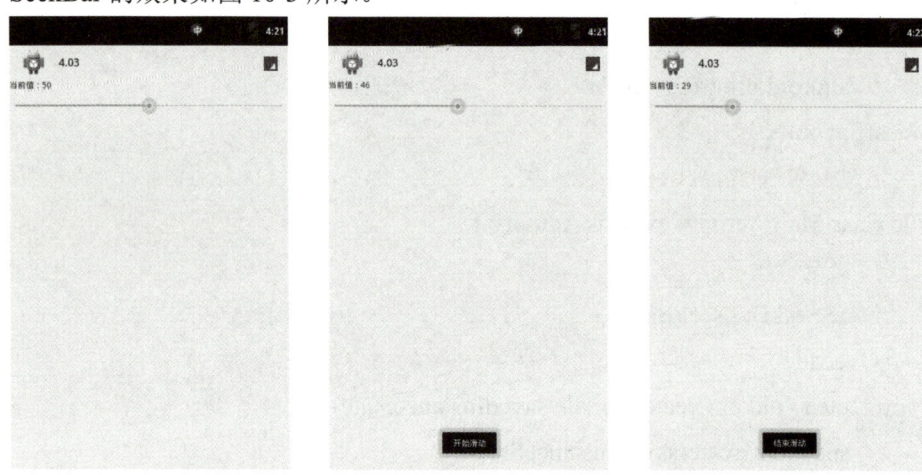

图 10-3　SeekBar 实现效果图

4. 星级评分条 RatingBar

星级评分条表示对某一事物或应用的支持、评价，或对某种应用服务的满意程度等，在星级评分条开发时，允许程序员直接通过拖动星星图案来改变进度。星级评分条与拖动条有着相同的一个父类 AbsSeekBar，因此用法、功能也十分接近，都允许用户通过拖动来改变进度。为了让程序能够响应星级评分条评分的改变，可以为它绑定一个 OnRatingBarChangeListener 监听器。

星级评分条支持的常见 XML 属性，如表 10-3 所示。

表 10-3　星级评分条支持的常见 XML 属性

XML 属性	说　明
android:isIndicator	设置该星级评分条是否允许用户改变（true 为不允许）
android:numStars	设置该星级评分条总共有多少个星
android:rating	设置该星级评分条默认的星级
android:stepSize	设置每次最少需要改变多少星级，默认 0.5 个

星级评分条常用方法有以下三个：
- getRating()：用于获取等级，表示选中了几颗星。
- getStepSize()：用于获取每次最少要改变多少个星级。
- getProgress()：用于获取进度，获取到的进度值= getRating()*getStepSize()。

下面是一个实现星级评分条的程序，代码如下：

1）布局文件 activity_main.xml

```xml
<LinearLayout xmlns:android="http://schemas.android.com/apk/res/android"
    android:layout_width="match_parent"
    android:layout_height="match_parent"
    android:orientation="vertical" >
    <!-- 星级评分条 -->
    <RatingBar
        android:id="@+id/ratingBar1"
        android:layout_width="wrap_content"
        android:layout_height="wrap_content"
        android:isIndicator="true"// true 不允许用户改变星级，false 允许用户改变星级
        android:numStars="5"     //设置星级个数
        android:rating="3.5" />  //设置初始星级
    <Button
        android:id="@+id/button1"
        android:layout_width="wrap_content"
        android:layout_height="wrap_content"
        android:text="提交" />
</LinearLayout>
```

2）主活动文件 MainActivity.Java

```java
public class MainActivity extends Activity {
// 成员变量
    private RatingBar ratingBar;                // 星级评分条
@Override
protected void onCreate(Bundle savedInstanceState) {
    super.onCreate(savedInstanceState);
    setContentView(R.layout.activity_main);
    // 获取星级评分条
```

```
ratingBar = (RatingBar) findViewById(R.id.ratingBar1);
// 获取按钮，并添加单击事件监听器
Button button = (Button) findViewById(R.id.button1);
button.setOnClickListener(new OnClickListener() {
@Override
public void onClick(View arg0) {
    int result = ratingBar.getProgress();        // 获取进度
    float rating = ratingBar.getRating();        // 获取等级
    float step = ratingBar.getStepSize();        // 获取每次最少改变多少颗星
    Log.i("星级评分条", "step=" + step + " result=" + result+ " rating=" + rating);
    Toast.makeText(MainActivity.this, "你得到了" + rating + "颗星",
    Toast.LENGTH_SHORT).show();
    }
});
}
}
```

实现效果如图 10-4 所示。

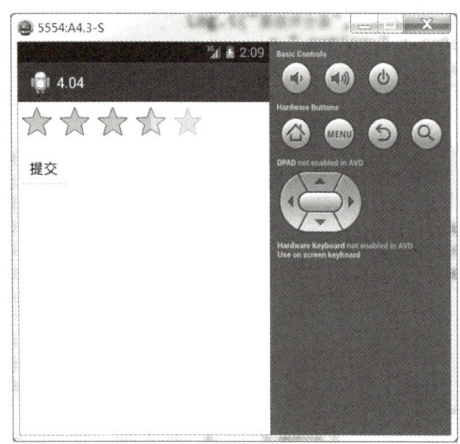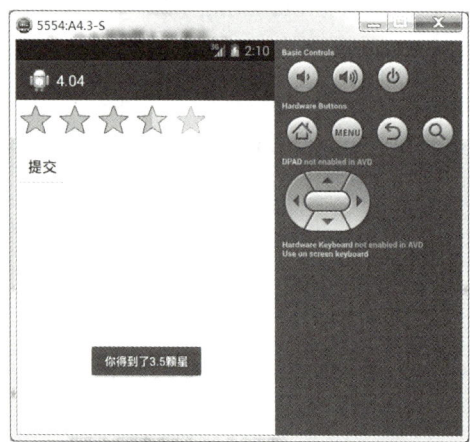

图 10-4　星形效果图实现

5. 选项卡 TabHost

选项卡（TabHost）主要功能是进行分类管理，它是一种非常实用的组件，可以在一个窗口中显示多组标签页，每个标签页 Tab 相当于获得了一个与外部容器相同大小的组件摆放区域。通过这种方式，就可以在一个容器里放置更多组件，如手机中的通话记录、未接来电、已接电话等。选项卡主要由 TabHost、TabWidget 和 FrameLayout 三个组件组成，如果程序需要监控 TabHost 里当前标签页的改变，可为它注册监听器 TabHost.OnTabChangeListener。

选项卡的实现一般按照如下步骤进行：

（1）在布局文件中添加所需的 TabHost、TabWidget 和 FrameLayout 组件。

（2）编写各标签页所对应的 XML 布局文件。

（3）在 Activity 中，获取并初始化 TabHost 组件。

（4）为 TabHost 对象添加标签页。

以下是选项卡的常用方法：

➢ newTabSpace(String tag)：创建选项卡。

➢ addTab(TabHost.TabSpec tabSpace)：添加标签页。

➢ getCurrentView()：获取当前的 View 组件。

➢ setup()：建立 TabHost 对象。

➢ setCurrentTab(int index)：设置当前显示的 Tab 编号。

➢ setCurrentTabByTag(String tag)：设置当前显示的 Tab 名称。

➢ getTabContentView()：返回标签容器 FrameLayout 的对象。

➢ setOnTabChangeListener(TabHost.OnTabChangeListener)：设置标签改变时触发。

下面是一个实现显示未接来电和已接来电选项卡的示例，程序如下：

1）布局文件 activity_main.xml

```xml
<TabHost xmlns:android="http://schemas.android.com/apk/res/android"
    android:id="@+id/tabhost"
    android:layout_width="match_parent"
    android:layout_height="match_parent" >
    <LinearLayout
        android:layout_width="match_parent"
        android:layout_height="match_parent"
        android:orientation="vertical" >
        <TabWidget
            android:id="@android:id/tabs"              //必须用系统的 id 为组件指定 id 属性
            android:layout_width="match_parent"
            android:layout_height="wrap_content" />
        <FrameLayout
```

```
        android:id="@android:id/tabcontent"    //必须用系统的 id 为组件指定 id 属性
        android:layout_width="match_parent"
        android:layout_height="match_parent" />
    </LinearLayout>
</TabHost>
```

2）编辑标签页 1 布局文件 tab1.xml

```
<?xml version="1.0" encoding="UTF-8"?>
<LinearLayout xmlns:android="http://schemas.android.com/apk/rcs/android"
    android:id="@+id/linearLayout02"
    android:layout_width="wrap_content"
    android:layout_height="wrap_content"
    android:orientation="vertical" >
    <TextView
        android:layout_width="match_parent"
        android:layout_height="wrap_content"
        android:paddingLeft="5px"
        android:text="简约但不简单" />
    <TextView
        android:layout_width="match_parent"
        android:layout_height="wrap_content"
        android:paddingLeft="5px"
        android:text="风铃草" />
</LinearLayout>
```

3）编辑标签页 2 布局文件 tab2.xml

```
<?xml version="1.0" encoding="UTF-8"?>
<LinearLayout xmlns:android="http://schemas.android.com/apk/res/android"
    android:id="@+id/linearLayout03"
    android:layout_width="wrap_content"
    android:layout_height="wrap_content"
    android:orientation="vertical" >
    <TextView
        android:layout_width="match_parent"
        android:layout_height="wrap_content"
```

```xml
        android:paddingLeft="5px"
        android:text="小草" />
    <TextView
        android:layout_width="match_parent"
        android:layout_height="wrap_content"
        android:paddingLeft="5px"
        android:text="淙儿" />
    <TextView
        android:layout_width="match_parent"
        android:layout_height="wrap_content"
        android:paddingLeft="5px"
        android:text="贺儿" />
</LinearLayout>
```

4）主活动文件 MainActivity.Java

```java
public class MainActivity extends Activity {
// 成员变量
    private TabHost tabHost;
    @Override
    protected void onCreate(Bundle savedInstanceState) {
        super.onCreate(savedInstanceState);
        setContentView(R.layout.activity_main);
        // 获取 TabHost 组件，并初始化
        tabHost = (TabHost) findViewById(R.id.tabhost);
        tabHost.setup();           // 初始化 TabHost 组件
        // 为 TabHost 对象添加标签页：一个用于模拟显示未接来电，
        //另一个用于模拟显示已接来电
        LayoutInflater inflater = LayoutInflater.from(this);
        //声明并实例化一个 LayoutInflater 对象
        inflater.inflate(R.layout.tab1, tabHost.getTabContentView());
        inflater.inflate(R.layout.tab2, tabHost.getTabContentView());
        tabHost.addTab(tabHost.newTabSpec("tab01").setIndicator("未接来电")
                .setContent(R.id.linearLayout02));        //添加第 1 个标签页
        tabHost.addTab(tabHost.newTabSpec("tab02").setIndicator("已接来电")
```

```
                    .setContent(R.id.linearLayout03));        //添加第 2 个标签页
        }
    @Override
    public boolean onCreateOptionsMenu(Menu menu) {
        // Inflate the menu; this adds items to the action bar if it is present.
        getMenuInflater().inflate(R.menu.main, menu);
        return true;
        }
    }
```

实现效果如图 10-5 所示。

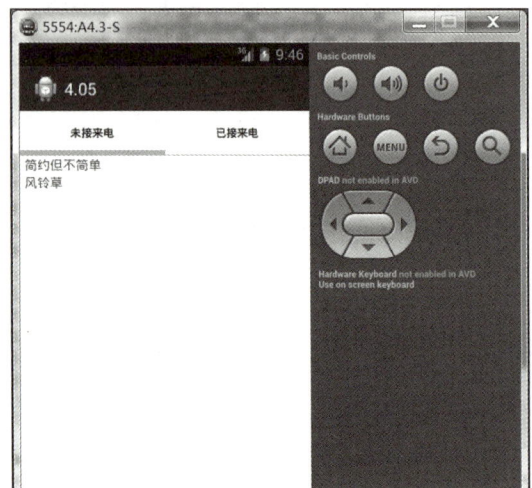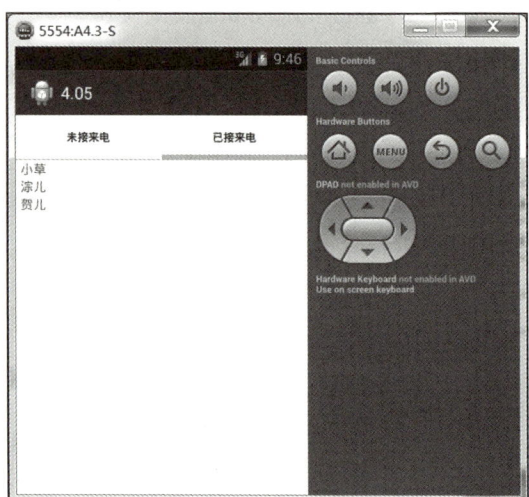

图 10-5　显示未接来电和已接来电的选项卡

6. 图像切换器 ImageSwitcher

图像切换器 ImageSwitcher 的主要功能是完成图片的切换显示，即可以单击按钮逐张切换显示的图片。在切换时，还可以为其添加一些动画效果。ImageSwitcher 是 ViewSwitcher 的子类。ViewSwitcher 类专门用于显示切换操作。在使用 ImageSwitcher 切换图片时，定义的 Activity 类必须实现 ViewSwitcher.ViewFactory 接口（视图切换工厂），并通过 makeView() 方法来创建用于显示图片的 ImageView。在使用 ImageSwitcher 切换图片时，可以通过 Animation 类指定切换图片时的动画显示效果。要想使用 Animation 类（动画类），还需使用 AnimationUtils 类完成。

图像切换器 ImageSwitcher 类常用的操作方法如下：

（1）ImageSwitcher(Context Context)：创建 ImageSwitcher 对象。

（2）setFactory(ViewSwitcher. ViewFactory factory)：设置 ViewFactory 对象，用于完成两个图片切换时 ViewSwitcher 的转换操作。

（3）setImageRecource(int resid)：设置显示的图片资源 ID。

（4）setInAnimation(Animation inAnimation)：图片读取进 ImageSwitcher 时的动画效果。

（5）setOutAnimation(Animation outAnimation)：图从 ImageSwitcher 消失时的动画效 AnimationUtils 类的主方法是 loadAnimation(Context context,int id)，通过该方法可以创建一个 Animation 对象。其中 id 为资源类型，可以直接从 android.R 类定义的常量中找出。（如 fade_in、fade_out，进入或离开时动画显示）

下面实现一个简单图片切换器，程序如下：

1）布局文件 activity_main.xml

```xml
<LinearLayout xmlns:android="http://schemas.android.com/apk/res/android"
    android:id="@+id/llayout"
    android:layout_width="match_parent"
    android:layout_height="match_parent"
    android:gravity="center"
    android:orientation="vertical"
    android:paddingTop="10px" >
    <!-- 图片切换器 -->
    <ImageSwitcher
        android:id="@+id/imageSwitcher1"
        android:layout_width="wrap_content"
        android:layout_height="wrap_content"
        android:layout_gravity="center" >
    </ImageSwitcher>
    <LinearLayout
        android:layout_width="match_parent"
        android:layout_height="match_parent"
        android:gravity="center"
        android:orientation="horizontal" >
        <Button
            android:id="@+id/button1"
```

```xml
            android:layout_width="wrap_content"
            android:layout_height="wrap_content"
            android:text="上一张" />
        <Button
            android:id="@+id/button2"
            android:layout_width="wrap_content"
            android:layout_height="wrap_content"
            android:text="下一张" />
    </LinearLayout>
</LinearLayout>
```

2）主活动文件 MainActivity.Java

```java
public class MainActivity extends Activity {
    /** 成员变量：共 3 个 */
    // 成员变量 1：保存要显示的图片 id 的数组
    int[] imageId = new int[] { R.drawable.flower01, R.drawable.flower02,
            R.drawable.flower03, R.drawable.flower04, R.drawable.flower05,
            R.drawable.flower06, R.drawable.flower07, R.drawable.flower08,
            R.drawable.flower09, R.drawable.flower10 };
    private int index = 0;                        //成员变量 2：保存当前图像索引
    private ImageSwitcher imageSwitcher;          //成员变量 3：图像切换器对象
    @Override
    protected void onCreate(Bundle savedInstanceState) {
        super.onCreate(savedInstanceState);
        setContentView(R.layout.activity_main);
        /**
         * 操作思路：
         * 1、获取图片切换器，为其设置淡入淡出动画效果
         * 2、为其设置 ViewSwitcher.ViewFactory,并实现 makeView 方法
         * 3、为图片切换器设置默认显示的图像
         */
        imageSwitcher = (ImageSwitcher) findViewById(R.id.imageSwitcher1);
        // 设置动画效果
        imageSwitcher.setInAnimation(AnimationUtils.loadAnimation(this,
```

```
        android.R.anim.fade_in));                // 设置淡入动画
        imageSwitcher.setOutAnimation(AnimationUtils.loadAnimation(this,
        android.R.anim.fade_out));               // 设置淡出动画
        //为其设置 ViewSwitcher.ViewFactory，并实现 makeView 方法
        imageSwitcher.setFactory(new ViewFactory() {
        @Override
        public View makeView() {
        // 实例化一个 ImageView 对象
        ImageView imageView = new ImageView(MainActivity.this);
        // 设置图片保持横纵比居中缩放
        imageView.setScaleType(ImageView.ScaleType.FIT_CENTER);
        // 设置图片布局参数
        imageView.setLayoutParams(new ImageSwitcher.LayoutParams(
                LayoutParams.WRAP_CONTENT, LayoutParams.WRAP_CONTENT));
        return imageView;               // 返回 ImageView 对象
        }
});
        imageSwitcher.setImageResource(imageId[index]);   //返回默认图片
        // 获取上一张、下一张图片，并为其设置单击事件监听器
        Button up = (Button) findViewById(R.id.button1);
        Button down = (Button) findViewById(R.id.button2);
        up.setOnClickListener(new OnClickListener() {
        @Override
        public void onClick(View arg0) {
            if (index > 0) {
                    index--;
            } else {
                    index = imageId.length - 1;
            }
        imageSwitcher.setImageResource(imageId[index]);   // 显示当前图片
        }
});
        down.setOnClickListener(new OnClickListener() {
```

```
            @Override
            public void onClick(View arg0) {
                    if (index < imageId.length - 1) {
                    index++;
                    } else {
                    index = 0;
                    }
                    imageSwitcher.setImageResource(imageId[index]);   //显示当前图片
            }
        });
    }
}
```

实现效果如图 10-6 所示。

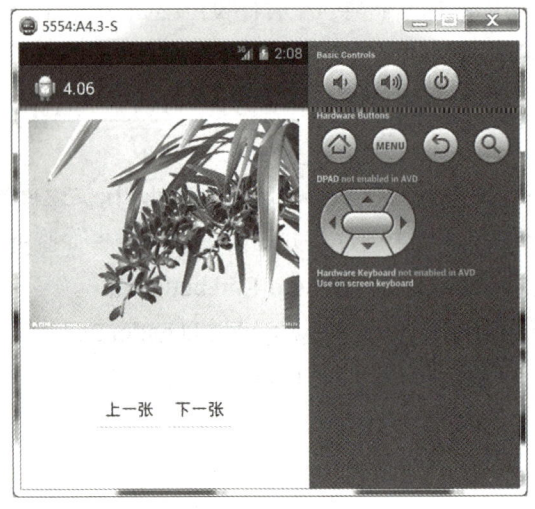

图 10-6　简单图片切换器效果图

7. 画廊视图 Gallery

画廊 Gallery 能够水平方向显示内容，并且可用手指直接拖动图片移动，一般用来浏览图片，被选中的选项位于中间，并且可用影响事件显示信息。在使用画廊视图时，首先在屏幕上添加 Gallery 组件。使用画廊视图，也需要使用 Adapter 提供显示的数据。通常使用 BaseAdapter 类为 Gallery 组件提供数据。

画廊支持的 XML 属性如表 10-4 所示。

表 10-4 Gallery 支持的 XML 属性

属　性	描　述
android:animationDuration	用于设置列表项切换的动画持续时间
android:gravity	用于设置对齐方式
android:spacing	用于设置列表项之间的间距
android:unselectedAlpha	用于设置没选中的列表项的透明度

下图是一个用于浏览图片的画廊，程序如下：

1）布局文件 activity_main.xml

```xml
<LinearLayout xmlns:android="http://schemas.android.com/apk/res/android"
    android:layout_width="match_parent"
    android:layout_height="match_parent">
    <!-- 画廊视图 -->
    <Gallery
    android:id="@+id/gallery1"
    android:layout_width="wrap_content"
    android:layout_height="wrap_content"
    android:unselectedAlpha="0.6"          //设置未选中时的透明度
    android:spacing="5px" />                //设置列表项间的间距
</LinearLayout>
```

2）属性资源文件 attrs.xml

```xml
<?xml version="1.0" encoding="utf-8"?>
<resources>
<declare-styleable name="Gallery">
<attr name="android:galleryItemBackground" />
</declare-styleable>
</resources>
```

3）主活动文件 MainActivity.Java

```java
public class MainActivity extends Activity {
    // 成员变量：定义并初始化图片 id 数组
        int[] imageId = new int[] { R.drawable.flower20, R.drawable.flower21,
        R.drawable.flower22, R.drawable.flower23, R.drawable.flower24,
        R.drawable.flower25, R.drawable.flower26, R.drawable.flower27,
        R.drawable.flower28, R.drawable.flower29, R.drawable.flower30,
```

```java
        R.drawable.flower31 };
@Override
protected void onCreate(Bundle savedInstanceState) {
    super.onCreate(savedInstanceState);
    setContentView(R.layout.activity_main);
    // 获取画廊视图
    Gallery gallery = (Gallery) findViewById(R.id.gallery1);
    // 创建适配器，来显示列表项内容
    BaseAdapter adapter = new BaseAdapter() {
        // 获取图片数量
        @Override
        public int getCount() {
            return imageId.length;
        }
        // 获取当前选项
        @Override
        public Object getItem(int arg0) {
            return arg0;
        }
        // 获取当前选项的 ID
        @Override
        public long getItemId(int arg0) {
            return arg0;
        }
        // 获取视图组件
        @Override
        public View getView(int arg0, View arg1, ViewGroup arg2) {
            ImageView imageView;
            if (arg1 == null) {
                imageView = new ImageView(MainActivity.this);
                imageView.setScaleType(ImageView.ScaleType.FIT_XY);
                imageView.setLayoutParams(new Gallery.LayoutParams(180, 135));
                TypedArray typedArray = obtainStyledAttributes(R.styleable.Gallery);
                imageView.setBackgroundResource(typedArray.getResourceId(
```

```
                    R.styleable.Gallery_android_galleryItemBackground,0));
            imageView.setPadding(5, 0, 5, 0);
        } else {
            imageView = (ImageView) arg1;
        }
        imageView.setImageResource(imageId[arg0]);
        return imageView;
    }
};
gallery.setAdapter(adapter);                    // 将适配器与 GridView 组件关联
gallery.setSelection(imageId.length / 2);       // 让中间的图片选中
// 为画廊视图添加项目单击事件监听器
gallery.setOnItemClickListener(new OnItemClickListener() {
    @Override
    public void onItemClick(AdapterView<?> arg0, View arg1, int arg2, long arg3)
    {
        Toast.makeText(MainActivity.this,"您选择了第" + String.valueOf(arg2 + 1) +
                "张图片",Toast.LENGTH_SHORT).show();
    }
});
}
```

画廊的实现效果如图 10-7 所示。

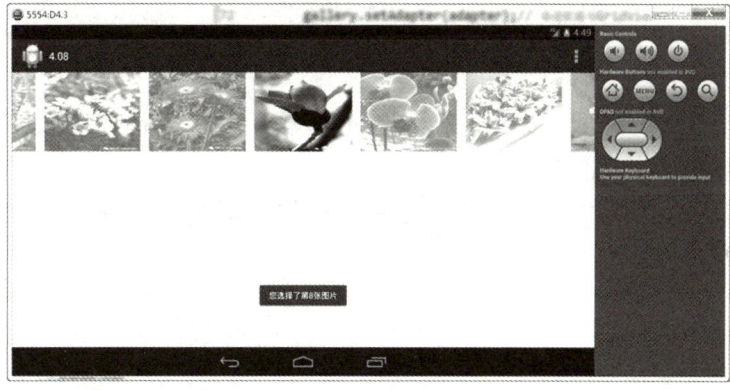

图 10-7　画廊

8. 消息提示框 Toast

在 Android 系统中经常会给用户显示一些提示信息，消息提示框被 Android 定义为 Toast，通常用于显示一些快速提示信息。Toast 类用于在屏幕中显示一个消息提示框，该消息提示框没有任何控制按钮，并且不会获得焦点，经过一段时间后自动消失。

使用消息提示框 Toast 类的步骤如下：

1）创建 Toast 类对象

创建 Toast 类对象的方法有两种，分别是：

- 使用构造方法（Toast toast=new Toast(this);）。
- 调用 makeText()方法创建一个名为 toast 的 Toast 对象，例如：Toast toast= Toast.makeText(this,"要显示的内容",TOAST.LENGTH_SHORT)，其中在参数中，this 表示在当前 Activity 类中被使用，TOAST.LENGTH_SHORT 代表短时显示。

2）调用 Toast 类提供的方法

经常使用的方法如表 10-5 所示。

表 10-5 Toast 支持的 XML 属性

方 法	描 述
setDuration(int duration)	用于设置消息提示框持续的时间，参数值通常使用 TOAST.LENGTH_SHORT 或 TOAST.LENGTH_LONG
setGravity(int gravity, int xOffset, int yOffset)	用于设置消息提示框的位置，参数 gravity 用于指定对齐方式，xOffset 和 yOffset 指定偏移值
setMargin(float horizontalMargin, float verticalMargin)	用于设置消息提示框的页边距
setText(CharSequence s)	用于设置消息提示框的要显示的文本内容
setView(View view)	用于设置将要在消息提示框中显示的视图
show()	显示消息提示框

3）调用 Toast 类的 show()方法

显示消息提示框。需要注意的是必须调用该方法，否则设置的消息提示框不显示。下面实现一个消息提示框，程序如下：

```
Toast.makeText(this, "我是通过 makeText()方法创建的消息显示框！", 1).show();
```

消息框的效果如图 10-8 所示。

项目十 高级用户界面设计

图 10-8 消息框的效果图

三、项目实施过程

下面我们就通过做一个实现仿 Windows 7 图片预览窗格效果的项目,来对 Android 中高级组件进行综合的应用。

1. 创建工程

创建名为 AndroidCode10 的 Android 工程,包结"com.xdxy.ui",如图 10-9 所示。

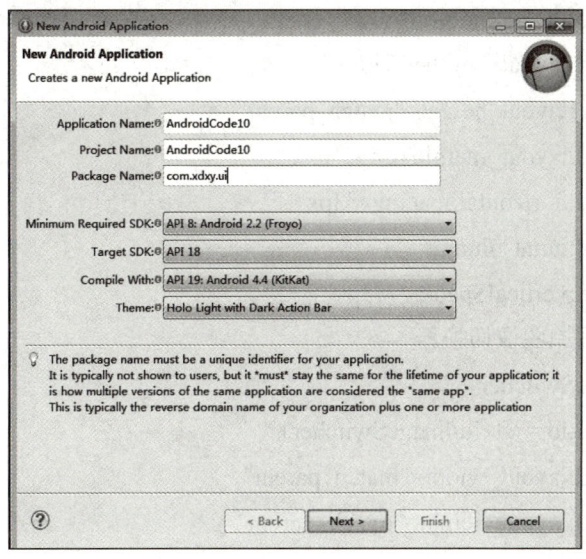

图 10-9 创建工程

2. XML 布局文件的开发

界面的设计效果如图 10-10 所示。

图 10-10　项目效果图

下面编辑布局文件 activity_main.xml，布局为 LinearLayout，方向为水平方向，并在其中添加，GridView 和 ImageSwitcher 两个组件。

```
<LinearLayout xmlns:android="http://schemas.android.com/apk/res/android"
android:layout_width="match_parent"
android:layout_height="match_parent"
android:orientation="horizontal" >
        <!-- 网格视图，4 列 -->
        <GridView
        android:id="@+id/gridView"
        android:layout_width="640px"
        android:layout_height="match_parent"
        android:layout_marginTop="10px"
        android:horizontalSpacing="3px"
        android:numColumns="4"
        android:verticalSpacing="3px" />
        <!-- 添加图像切换器 -->
        <ImageSwitcher
        android:id="@+id/imageSwitcher1"
        android:layout_width="match_parent"
        android:layout_height="match_parent"
        android:padding="30px" >
```

```
            </ImageSwitcher>
</LinearLayout>
```

3. Java 文件的开发

（1）编辑主活动文件 MainActivity.Java，实现 GridView 和 ImageSwitcher。

```java
public class MainActivity extends Activity {
    // 成员变量
        private int[] imageId = new int[] { R.drawable.flower01,
            R.drawable.flower02, R.drawable.flower03, R.drawable.flower04,
            R.drawable.flower05, R.drawable.flower06, R.drawable.flower07,
            R.drawable.flower08, R.drawable.flower09, R.drawable.flower10,
            R.drawable.flower11, R.drawable.flower12 };
        private ImageSwitcher imageSwitcher;
        @Override
    protected void onCreate(Bundle savedInstanceState) {
        super.onCreate(savedInstanceState);
        setContentView(R.layout.activity_main);
        // 获取图像切换器，并为其设置淡出淡入效果
        // 然后为其设置一个 ImageSwitcher.ViewFactory,并重写 makeView()方法
        // 最后为图像切换器设置默认显示的图像
        imageSwitcher = (ImageSwitcher) findViewById(R.id.imageSwitcher1);
        imageSwitcher.setInAnimation(AnimationUtils.loadAnimation(this,
            android.R.anim.fade_in));
        imageSwitcher.setOutAnimation(AnimationUtils.loadAnimation(this,
            android.R.anim.fade_out));
        imageSwitcher.setFactory(new ViewFactory() {
            @Override
            public View makeView() {
                // 实例化一个 ImageView 组件
                ImageView imageView = new ImageView(MainActivity.this);
                // 设置保持横纵比居中缩放图像
                imageView.setScaleType(ImageView.ScaleType.FIT_CENTER);
                // 设置图像的布局参数
```

```java
            imageView.setLayoutParams(new ImageSwitcher.LayoutParams(
                    LayoutParams.WRAP_CONTENT, LayoutParams.WRAP_CONTENT));
            return imageView;
        }
    });
    // 设置默认显示的图像
    imageSwitcher.setImageResource(imageId[0]);
    // 获取网格视图组件
    GridView gridView = (GridView) findViewById(R.id.gridView);
    // 创建 BaseAdapter 适基础配器，重写其中的方法
    BaseAdapter adapter = new BaseAdapter() {
        /**
         * 功能：获取图片数量
         */
        @Override
        public int getCount() {
            return imageId.length;
        }
        /**
         * 功能：获取当前选项
         */
        @Override
        public Object getItem(int arg0) {
            return arg0;
        }
        /**
         * 功能：获取当前选项的 id
         */
        @Override
        public long getItemId(int arg0) {
            return arg0;
        }
        /**
```

* 功能：获取视图组件
*/
@Override
public View getView(int arg0, View arg1, ViewGroup arg2) {
ImageView imageView;
if (arg1 == null) {
    // 实例化图像视图
    imageView = new ImageView(MainActivity.this);
    /***** 设置图片的宽度、高度、内边距 ******/
    imageView.setAdjustViewBounds(true);
    imageView.setMaxWidth(150);
    imageView.setMaxHeight(113);
    imageView.setPadding(5, 5, 5, 5);
} else {
        imageView = (ImageView) arg1;
}
// 设置 imageView 要显示想图片
imageView.setImageResource(imageId[arg0]);
return imageView;
}
};
//设置适配器与 GridView 的关联，并设置单击图片显示事件监听器
    gridView.setAdapter(adapter);
        gridView.setOnItemClickListener(new OnItemClickListener(){
        @Override
        public void onItemClick(AdapterView<?> arg0, View arg1, int arg2,
        long arg3) {
            //显示选中的图片
            imageSwitcher.setImageResource(imageId[arg2]);
        }});
    }
}

（2）程序编写好后，将其运行到 Android 开发终端上进行测试。

四、项目思考与扩展

1. 实现如图 10-11 所示效果,拖动星级能够改变图片的透明度。

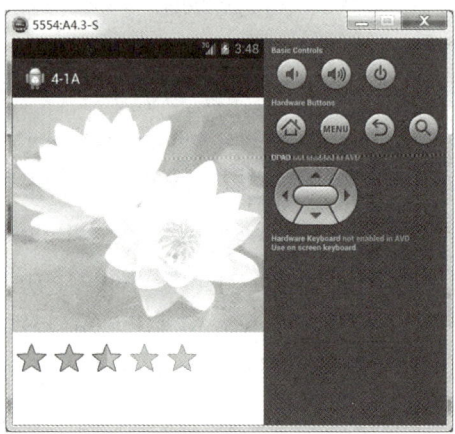

图 10-11　拖动星级能够改变图片的透明度

2. 实现如图 10-12 所示幻灯片式图片浏览器程序的开发。

图 10-12　幻灯片式图片浏览器

项目十一 列表视图 ListView

【本章导读】

在 Android 程序设计开发中，程序员有时候需要为用户显示许多数据，为了将这些数据整齐地以列表的形式显示在屏幕上，就需要使用 ListView 这个控件，本章项目我们就来学习如何应用 ListView 进行数据显示。

一、项目要求

1. 掌握 Activity 直接继承 ListActivity 实现 ListView 的方法。
2. 掌握使用适配器 Adapter 实现 ListView 的方法。
3. 实现带图片的 ListView 项目。

二、项目相关知识

1. ListView

ListView（列表视图）是 android 中使用非常广泛的组件之一，它以垂直列表的形式列出需要显示的列表项。向屏幕中添加 ListView 有以下两种方法：

➤ **方法 1**：在布局文件中，直接使用 ListView 组件创建，然后为 ListView 设置需要显示的列表内容（Adapter）。

➤ **方法 2**：让 Activity 直接继承 ListActivity 实现，然后通过 ListActivity 的 setListAdapter() 方法设置需要显示的列表内容。

ListView 支持的常用属性如表 11-1 所示。

表 11-1 ListView 支持的 XML 属性

属 性	描 述
android:divider	设置分割条，既可以用颜色分割，又可以用 Drawable 资源分割
android:dividerHeight	设置分割条的高
android:entries	用于通过数值资源为 ListView 指定列表项
android:footerDividersEnabled	用于设置是否在 footer view 之前设置分割条，默认 true，false 表示不绘制。使用该属性时，需要先通过 addFooterView()方法为 ListView 设置 footer view
android:headerDividersEnabled	用于设置是否在 header view 之后设置分割条，默认 true，false 表示不绘制。使用该属性时，需要先通过 addHeaderView()方法为 ListView 设置 header view

如果想在单击 ListView 的各列表项时，获取选择项的值，需要为 ListView 添加 OnItimClickListener 事件监听器。

2. 直接使用 ListView 组件创建 ListView

使用 ListView 进行列表视图的开发过程如下：

1）创建数组资源文件 arrays.xml

```xml
<?xml version="1.0" encoding="UTF-8"?>
<resources>
    <string-array name="ctype">
        <item>情景模式</item>
        <item>主题模式</item>
        <item>手机</item>
        <item>程序管理</item>
        <item>通话设置</item>
        <item>连接功能</item>
    </string-array>
</resources>
```

2）布局文件 activity_main.xml

```xml
<LinearLayout xmlns:android="http://schemas.android.com/apk/res/android"
    android:layout_width="match_parent"
    android:layout_height="match_parent"
    >
```

```
<ListView//列表视图组件
    android:id="@+id/ListView1"
    android:layout_width="match_parent"
    android:layout_height="wrap_content"
    android:divider="@drawable/greendivider"//设置分隔条：指定分割条采用图片
    android:dividerHeight="3dp"//设置分隔条高度
    android:footerDividersEnabled="false"//设置在列表视图底部不绘制分隔条
    android:headerDividersEnabled="false" />//设置在列表视图顶部不绘制分隔条
</LinearLayout>
```

3）主活动 MainActivity.Java

```
public class MainActivity extends Activity {
@Override
    protected void onCreate(Bundle savedInstanceState) {
        super.onCreate(savedInstanceState);
        setContentView(R.layout.activity_main);
        // 获取 ListView 组件
        final ListViewListView = (ListView) findViewById(R.id.ListView1);
        // 为 ListView 组件，设置 header view
        ListView.addHeaderView(line());
        // 采用指定的图像视图，绘制 header view
        ArrayAdapter<CharSequence> adapter = ArrayAdapter.createFromResource(
                this, R.array.ctype, android.R.layout.simple_list_item_checked);
        // 为 ListView 组件，设置列表项
        ListView.setAdapter(adapter);
        // 将适配器与 ListView 关联
        ListView.addFooterView(line());
        // 为 ListView 组件，设置 footer view
        ListView.setOnItemClickListener(new OnItemClickListener() {
        // 为 ListView 组件，设置单击选择项事件监听器
        @Override
        public void onItemClick(AdapterView<?> parent, View arg1, int pos,long id) {
            // 获取选择项的值，并显示提示框
            String result = parent.getItemAtPosition(pos).toString();
```

```
            Toast.makeText(MainActivity.this, result, Toast.LENGTH_SHORT).show();
        }
    });
}

// 成员方法：利用资源中的图像资源，来创建一个图像视图。
private View line() {
    ImageView image = new ImageView(this);
image.setImageResource(R.drawable.line1);
return image;
    }
}
```

实现效果如图 11-1 所示。

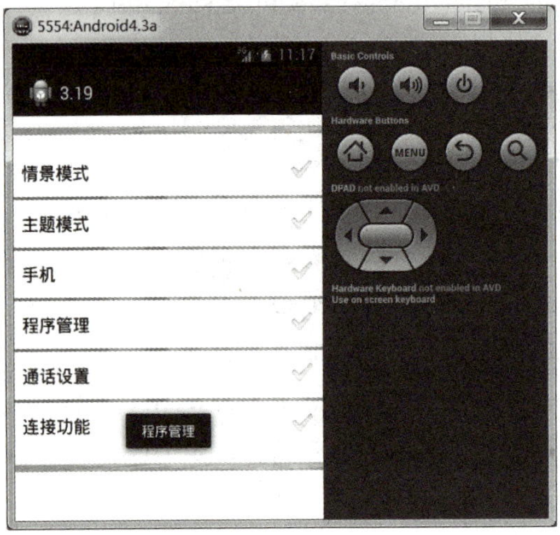

图 11-1　ListView 实现效果

3. 让 Activity 继承 ListActivity 实现列表

如果程序的窗口中仅仅需要一个列表，则可以直接让 Activity 继承 ListActivity 来实现。继承了 ListActivity 的类，无需调用 setContentVIew()方法来显示页面，可直接为其设置适配器，来显示一个列表。

其实现过程只需编写一个主活动文件 MainActivity.Java，程序如下：

```
public class MainActivity extends ListActivity {
    @Override
```

```java
    protected void onCreate(Bundle savedInstanceState) {
        super.onCreate(savedInstanceState);
// 在这个 ListActivity 中，不需要设置布局文件，下面的语句删除或添加注释
// setContentView(R.layout.activity_main);
// 创建适配器，用于指定列表项
String[] songs = new String[] { "哭砂", "时间煮雨", "有没有人告诉你", "栀子花开", "等待" };
ArrayAdapter<String> adapter =
new ArrayAdapter<String>(this,android.R.layout.simple_list_item_single_choice, songs);
// 在该窗口中，设置列表内容
this.setListAdapter(adapter);
}
// 重写 ListActivity 的 onListItemClick()方法，来获取列表的选择项值
@Override
protected void onListItemClick(ListView l, View v, int position, long id) {
    super.onListItemClick(l, v, position, id);
    String result = l.getItemAtPosition(position).toString();
    Toast.makeText(MainActivity.this, result, Toast.LENGTH_SHORT).show();
}
}
```

实现效果如图 11-2 所示。

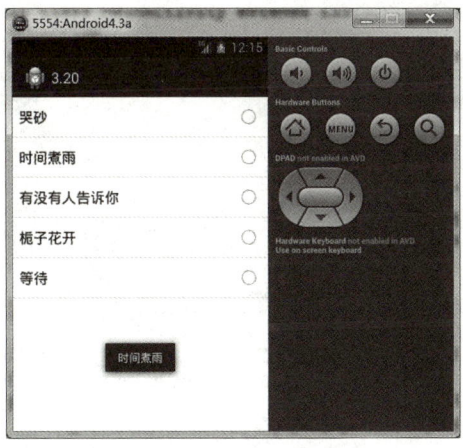

图 11-2　ListView 实现效果

三、项目实施过程

下面我们就通过做一个实现带图标的 ListView 项目，来进行 ListView 的实际使用。实现效果图如图 11-3 所示。

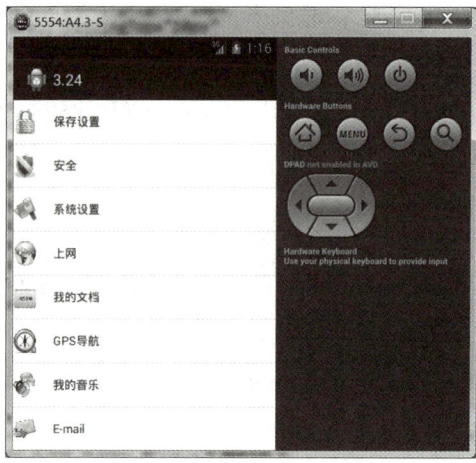

图 11-3　项目效果图

1. 创建工程

创建名为 AndroidCode11 的 Android 工程，包结构为"com.xdxy.ListView"，Activity 名为 MenuActivity，如图 11-4 所示。

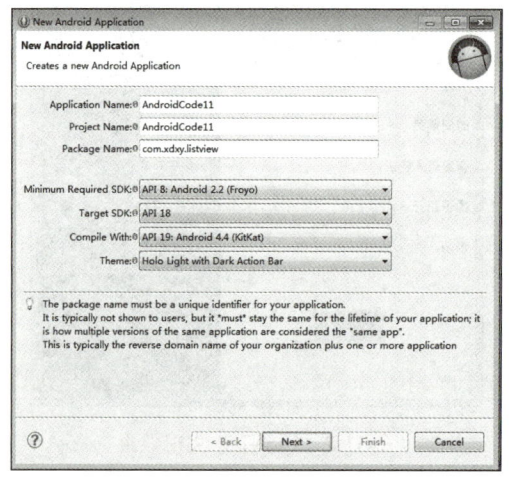

图 11-4　创建工程

2. XML 布局文件的开发

1）修改编辑文件 activity_main.xml，拖入 ListView 组件

```xml
<LinearLayout xmlns:android="http://schemas.android.com/apk/res/android"
    android:layout_width="match_parent"
    android:layout_height="match_parent" >
    <ListView
        android:id="@+id/ListView"
        android:layout_width="wrap_content"
        android:layout_height="wrap_content" />
</LinearLayout>
```

2）创建菜单项布局文件 items.xml

```xml
<?xml version="1.0" encoding="utf-8"?>
<LinearLayout
    xmlns:android="http://schemas.android.com/apk/res/android"
    android:orientation="horizontal"
    android:layout_width="match_parent"
    android:layout_height="match_parent">
    <ImageView
        android:id="@+id/image"
        android:paddingRight="10px"
        android:paddingTop="10px"
        android:paddingBottom="10px"
        android:adjustViewBounds="true"
        android:maxWidth="50px"
        android:maxHeight="50px"
        android:layout_height="wrap_content"
        android:layout_width="wrap_content"/>
    <TextView
        android:layout_width="wrap_content"
        android:layout_height="wrap_content"
        android:padding="10px"
        android:layout_gravity="center"
        android:id="@+id/title" />
```

</LinearLayout>

3. Java 文件的开发

(1) 修改主活动文件 MainActivity.Java。

```java
public class MainActivity extends Activity {
    @Override
    protected void onCreate(Bundle savedInstanceState) {
        super.onCreate(savedInstanceState);
        setContentView(R.layout.activity_main);
        // 获取 ListView 组件
        ListView ListView = (ListView) findViewById(R.id.ListView);
        // 定义一个保存图片的数组
        int[] imageId = new int[] { R.drawable.img01, R.drawable.img02,
                        R.drawable.img03, R.drawable.img04, R.drawable.img05,
                        R.drawable.img06, R.drawable.img07, R.drawable.img08 };
        // 定义一个保存文字的数组
        String[] title = new String[] { "保存设置", "安全", "系统设置", "上网", "我的文档",
                        "GPS 导航", "我的音乐", "E-mail" };
        // 创建一个 List 集合
        List<Map<String, Object>> listItems = new ArrayList<Map<String, Object>>();
        // 通过 for 循环将图片 id 和列表项文字放到 Map 中，并添加到 List 集合
        for (int i = 0; i < imageId.length; i++) {
            Map<String, Object> map = new HashMap<String, Object>();// 实例化 map 对象
            map.put("image", imageId[i]);
            map.put("title", title[i]);
            listItems.add(map);
        }
        SimpleAdapter adapter = new SimpleAdapter(this, listItems, R.layout.items,
                        new String[] { "title", "image" }, new int[] {R.id.title, R.id.image });
        ListView.setAdapter(adapter);
    }
    @Override
    public boolean onCreateOptionsMenu(Menu menu) {
```

// Inflate the menu; this adds items to the action bar if it is present.
getMenuInflater().inflate(R.menu.main, menu);
return true;
　　}
}

（2）保存程序，在模拟器上运行。

四、项目思考与扩展

实现带图片和文字的仿 QQ 效果的 ListView，如图 11-5 所示。

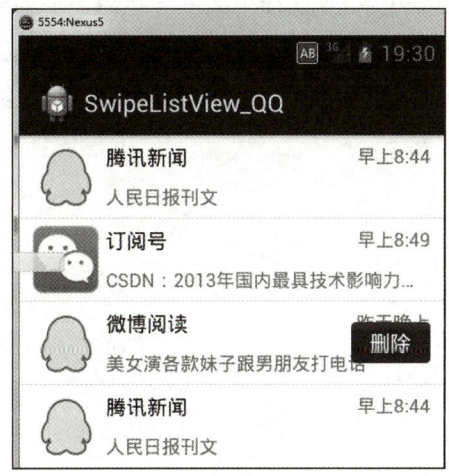

图 11-5　项目扩展效果图

项目十二　自定义菜单

【本章导读】

菜单是 Android 应用程序 UI 设计中必不可缺的一个部分，Android 系统中为我们提供了三种不同形式的菜单：可选项菜单（Options Menu）、子菜单（SubMenu）和上下文菜单（Context Menu）。开发人员可根据设计需要自行选择。本章，我们就学习菜单的设计与开发。

一、项目要求

1．掌握动态生成菜单的方法。
2．掌握将菜单配置到资源文件的方法。
3．掌握选项菜单的使用方法。
4．掌握子菜单的使用方法。
5．掌握快捷菜单的使用方法。

二、项目相关知识

菜单是应用程序中非常重要的组成部分，能够在不占用界面空间的前提下，为应用程序提供统一的选择功能和设置界面。Android 系统支持三种菜单模式，分别是选项菜单、子菜单和快捷菜单。

Android 程序的菜单可以在代码中动态生成，也可以使用 XML 文件制作菜单资源，然后通过 inflate() 函数映射到程序代码中。使用 XML 文件描述菜单是程序员开发时较好的选择，可以将菜单内容与代码分离，有利于分析和调整菜单结构。

- ➢ **选项菜单**：选项菜单是一种经常使用的菜单，用户可以通过菜单键打开选项菜单。
- ➢ **子菜单**：子菜单就是二级菜单，用户单击选项菜单或快捷菜单中的菜单项就可以打开子菜单。
- ➢ **快捷菜单**：快捷菜单类似于计算机程序中的"右键菜单"，当用户单击界面上某个元素超过 2 秒后，将启动注册到该界面元素的快捷菜单。

三、项目实施过程

下面我们就通过做一个简单的程序，来学习菜单的使用。

1. 创建工程

创建名为 AndroidCode12 的 Android 工程，包结构为"com.xdxy.menu"，Activity 名为 MenuActivity，如图 12-1 所示。

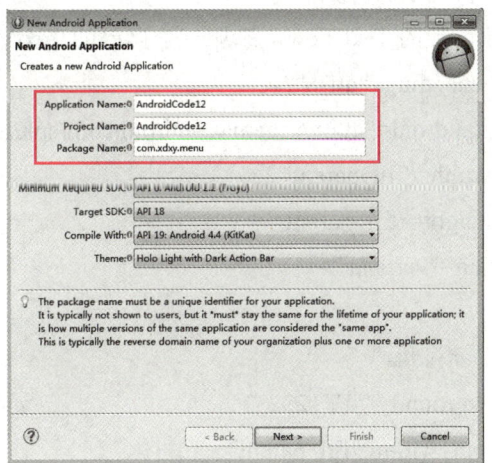

图 12-1　创建工程

2. XML 布局文件的开发

（1）设计初始界面，如图 12-2 所示。

图 12-2　项目效果图

（2）通过【res】→【layout】路径，找到 main.xml 文件，双击打开，然后单击下面的【main.xml】标签进入代码编辑界面。

（3）删除 main.xml 自带的 TextView，添加一个自己的 TextView。

```
<?xml version="1.0" encoding="utf-8"?>
<LinearLayout xmlns:android="http://schemas.android.com/apk/res/android"
    android:layout_width="fill_parent"
    android:layout_height="fill_parent"
    android:orientation="vertical" >
    <TextView
        android:id="@+id/tv"
        android:background="#FF123232"
        android:layout_width="fill_parent"
        android:layout_height="100dp"
        android:gravity="center_vertical|center_horizontal"
        android:textSize="30sp"
        android:text="长按 2 秒" />
</LinearLayout>
```

（4）按【Ctrl+S】组合键保存退出，右击 res 目录，在弹出的快捷菜单中单击【New】→【Folder】命令，如图 12-3 所示。

项目十二 自定义菜单

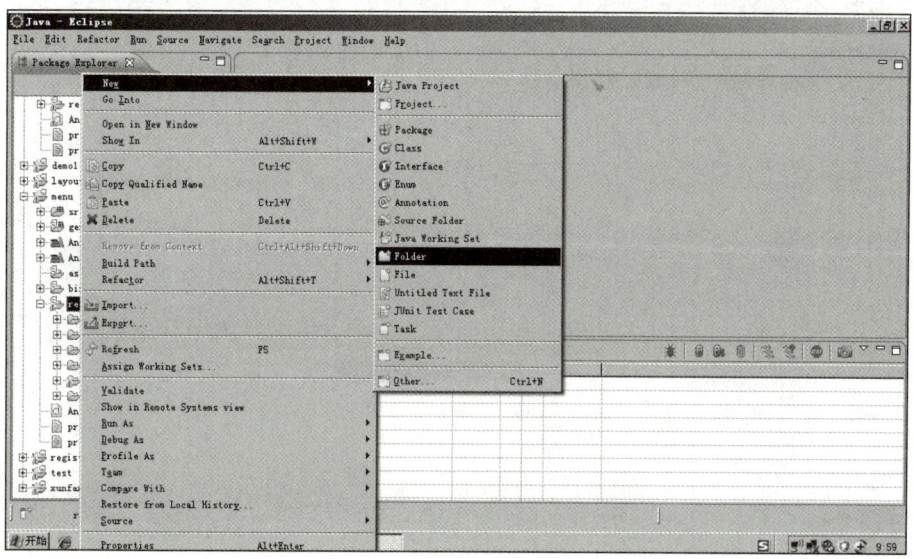

图 12-3 单击 Folder

（5）在 New Folder 对话框中输入 Folder name 为 menu，单击【Finish】按钮，如图 12-4 所示。

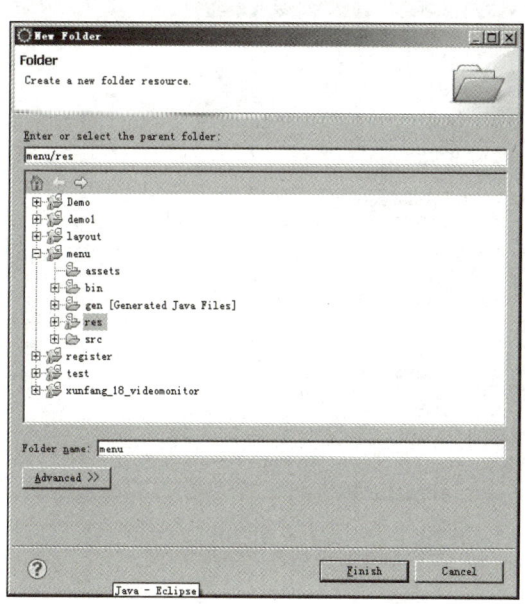

图 12-4 New Folder 对话框

（6）右击 res 目录下的 menu 文件夹，在弹出的快捷菜单中选择【New】→【Other】命令，如图 12-5 所示。

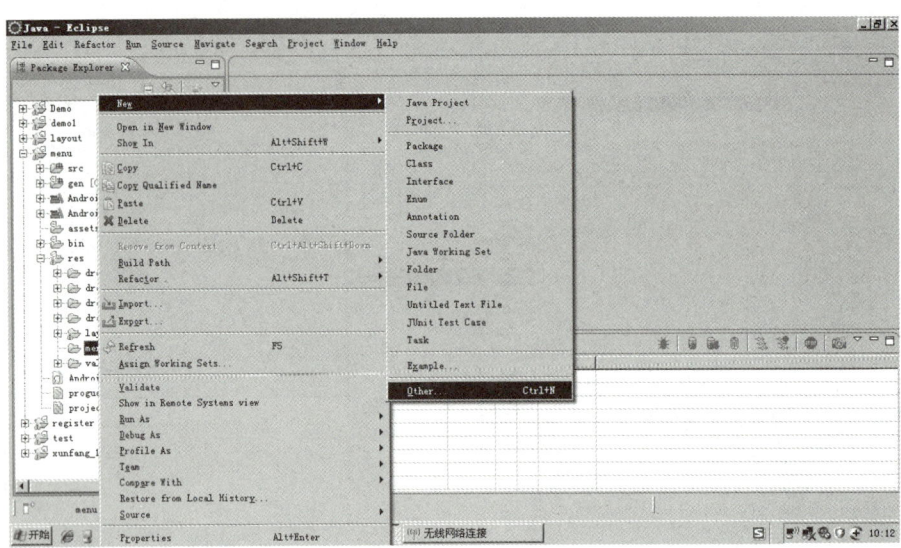

图 12-5 单击 Other

（7）在弹出的 New 对话框中，打开 XML 文件夹，选中里面的 XML File 项，然后单击【Next】按钮，如图 12-6 所示。

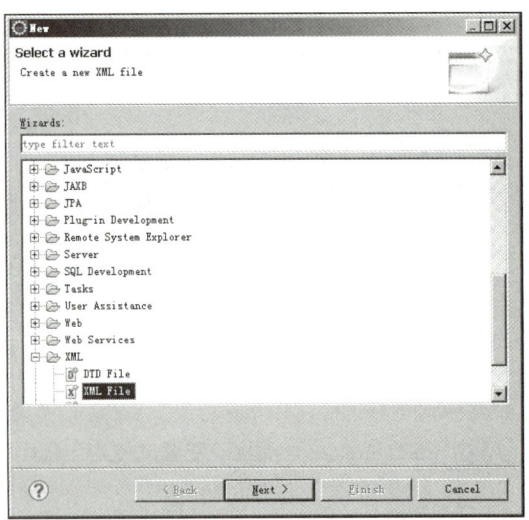

图 12-6 选中 XML File

（8）在 New XML File 对话框中，输入 File name 为 contextenu.xml，然后单击【Finish】按钮，如图 12-7 所示。

项目十二 自定义菜单

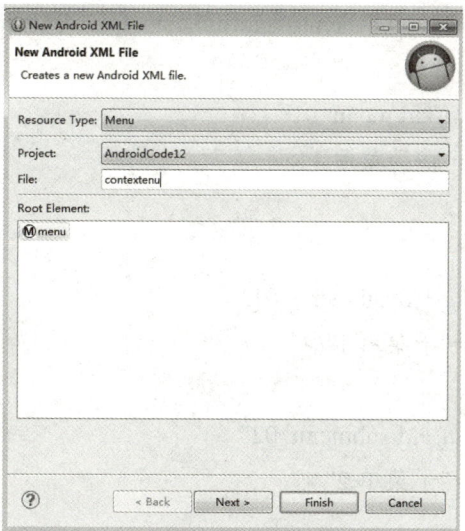

图 12-7 输入文件名

（9）双击 contextmenu.xml 文件打开，如图 12-8 所示。

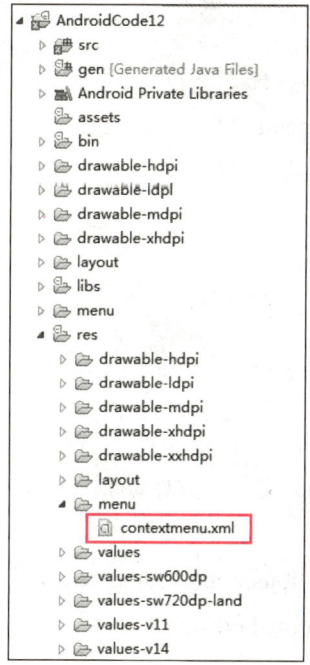

图 12-8 双击打开

（10）在 contextmenu.xml 文件中，添加代码，添加好后按【Ctrl+S】键保存退出。

<?xml version="1.0" encoding="UTF-8"?>

```xml
<menu xmlns:android="http://schemas.android.com/apk/res/android">
    <item
        android:id="@+id/menu_01"
        android:title="快捷菜单 1">
        <menu >
        <item
        android:id="@+id/submenu_01"
        android:title="子菜单 1"/>
        <item
        android:id="@+id/submenu_02"
        android:title="子菜单 2"/>
        <item
        android:id="@+id/submenu_03"
        android:title="子菜单 3"/>
        </menu>
        </item>
    <item
        android:id="@+id/menu_02"
        android:title="快捷菜单 2"/>
    <item
        android:id="@+id/menu_03"
        android:title="快捷菜单 3"/>
</menu>
```

3. Java 文件的开发

（1）打开 MenuActivity.Java 文件，开始编辑源代码，为界面添加选项菜单及菜单项单击事件，该菜单动态生成。

```java
public boolean onCreateOptionsMenu(Menu menu) {
    // TODO Auto-generated method stub
    menu.add(0,1,1,"菜单项 1");
    menu.add(0,2,2,"菜单项 2");
    menu.add(0,3,3,"菜单项 3");
    return true;
```

```java
}
@Override
public boolean onOptionsItemSelected(MenuItem item) {
    // TODO Auto-generated method stub
    switch(item.getItemId()){
    case 1:
        Toast.makeText(this,"您单击了菜单项1",2000).show();
        break;
    case 2:
        Toast.makeText(this,"您单击了菜单项2",2000).show();
        break;
    case 3:
        Toast.makeText(this,"您单击了菜单项3",2000).show();
        break;
    }
    return true;
}
```

（2）添加快捷菜单、快捷菜单的子菜单及菜单项的单击事件，该快捷菜单及其子菜单在 XML 文件中配置。

```java
public void onCreateContextMenu(ContextMenu menu, View v,
        ContextMenuInfo menuInfo) {
    MenuInflater inflater = getMenuInflater();
    inflater.inflate(R.menu.contextmenu, menu);
}
public boolean onContextItemSelected(MenuItem item) {
    switch(item.getItemId()){
    case R.id.menu_02:
        Toast.makeText(this,"您单击了快捷菜单项2",2000).show();
        break;
    case R.id.menu_03:
        Toast.makeText(this,"您单击了快捷菜单项3",2000).show();
    break;
    case R.id.submenu_01:
```

```
        Toast.makeText(this,"您单击了子菜单项 1",2000).show();
    break;
case R.id.submenu_02:
        Toast.makeText(this,"您单击了子菜单项 2",2000).show();
            break;
case R.id.submenu_03:
        Toast.makeText(this,"您单击了子菜单项 3",2000).show();
            break;
    }
    return true;
}
```

(3) 为界面控件 TextView 注册快捷菜单。

```
public void onCreate(Bundle savedInstanceState) {
        super.onCreate(savedInstanceState);
        setContentView(R.layout.main);
        TextView tv = (TextView)findViewById(R.id.tv);
        registerForContextMenu(tv);
    }
```

(4) 按【Ctrl+S】组合键保存，运行程序。部分效果如图 12-9 和图 12-10 所示。

初始效果

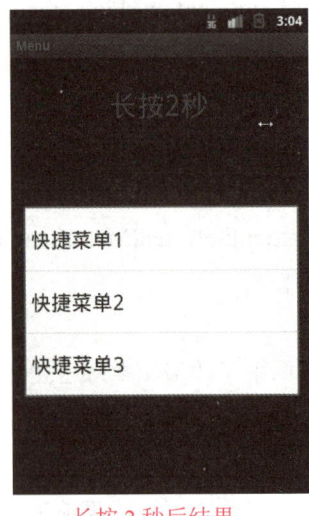
长按 2 秒后结果

单击【快捷菜单 2】结果

图 12-9 程序运行效果

项目十二　自定义菜单

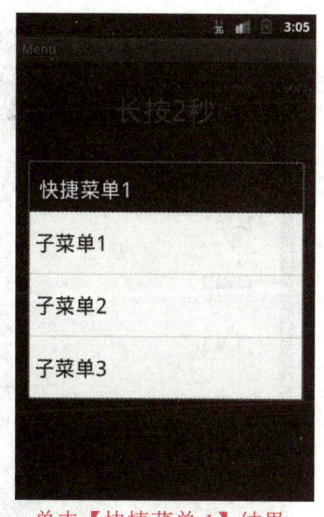

单击【快捷菜单 1】结果　　　　单击【子菜单 1】结果

图 12-10　程序运行效果

四、项目思考与扩展

修改程序，增加一个菜单选项，用两种方式实现：在代码中动态添加，在 XML 文件中添加，并增加一个子菜单项，如图 12-11 所示。

　　初始状态　　　　　　　　动态添加菜单选项

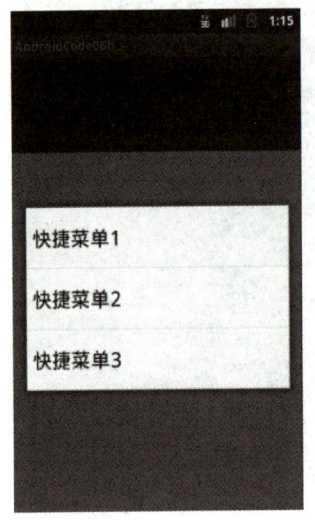
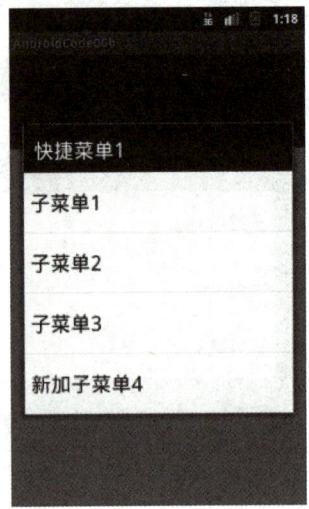

xml 添加子菜单

图 12-11　程序运行效果

项目十三　BroadCastReceiver 实现广播的接收与发送

【本章导读】

在 Android 系统中，经常需要处理地域变换、电量不足、来电来信等信息，这些都是 Android 的广播系统发起的。开发者也可以自行编写程序来播放一个广播，本章我们就来学习 Android 广播的接收与发送。

一、项目要求

1. 掌握在程序中动态注册和取消注册广播接收器的方法。
2. 掌握静态注册广播接收器的方法。
3. 掌握发送广播的方法。

二、项目相关知识

1. 广播

在 Android 中广播分为两个部分：一个部分是广播发送者，另外一个部分是广播接收者。一般来说，BroadcastReceiver 类指的就是广播接收者，也可以叫做广播的接收器。发送广播时，也就是通过 Intent 将要发送的信息发送出去的，发送广播的方法是 sendBroadcast()，参数是 Intent 类型，Intent 可以携带数据。广播作为 Android 系统内部各个组件之间的通信方式，应用的场景有以下几个方面：

（1）在 Android 内部同一组件之间进行消息的通信。

（2）在 Android 内部不同组件之间进行消息的通信。

（3）在 Android 内部不同线程之间进行消息的通信。

（4）在不同 Android 项目之间进行消息的传递和通信。

（5）Android 系统在某种特定情况下进行消息的传递和通信。

2. 静态和注册广播接收器

在 Android 的主配置文件中，通过<receiver>标签注册的广播接收器是静态注册广播接收器。广播接收器也可以在程序代码中动态地注册和取消，这种广播接收器称为动态注册广播接收器。

三、项目实施过程

下面我们通过一个程序来学习广播的使用。

1. 创建工程

创建名为 AndroidCode13 的 Android 工程，包结构为"com.xdxy.broadcast"，Activity 名为 IntentActivity，如图 13-1 所示。

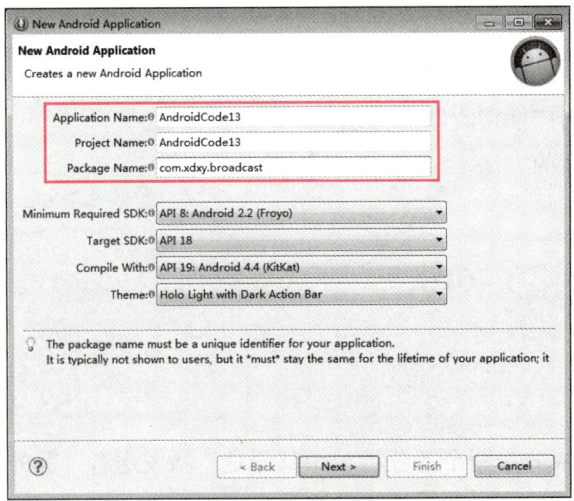

图 13-1　创建工程

2. XML 布局文件的开发

（1）设计初始界面，如图 13-2 所示。

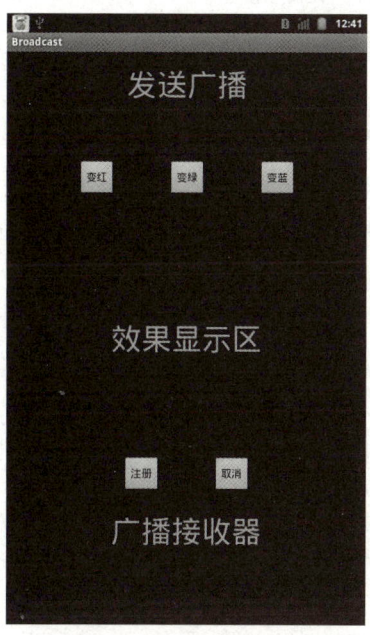

图 13-2　项目效果图

（2）在主配置文件中，注册一个广播接收器，并添加接收系统启动的广播。

```
<receiver android:name=".broadcast.MyReceiver">
    <intent-filter >
        <action android:name="android.intent.action.BOOT_COMPLETED"/>
    </intent-filter>
</receiver>
```

（3）为程序设计界面，界面分为 3 部分：上部发送广播、下部注册广播接收器和中部显示效果。

```
<?xml version="1.0" encoding="utf-8"?>
<LinearLayout xmlns:android="http://schemas.android.com/apk/res/android"
    android:layout_width="fill_parent"
    android:layout_height="fill_parent"
    android:orientation="vertical" >
    <LinearLayout
        android:layout_width="fill_parent"
```

```xml
        android:layout_height="0dp"
        android:layout_weight="1.0"
        android:orientation="vertical">

    <LinearLayout
        android:layout_width="fill_parent"
        android:layout_height="0dp"
        android:layout_weight="1.0">

        <TextView
            android:layout_width="fill_parent"
            android:layout_height="fill_parent"
            android:gravity="center_vertical|center_horizontal"
            android:textSize="40dp"
            android:text="发送广播"/>
</LinearLayout>
    <LinearLayout
        android:layout_width="fill_parent"
        android:layout_height="0dp"
        android:layout_weight="2.0"
        android:gravity="center_vertical|center_horizontal">

        <Button
            android:id="@+id/btToRed"
            android:layout_width="wrap_content"
            android:layout_height="wrap_content"
            android:text="变红"
            android:onClick="doClick"/>
        <Button
            android:id="@+id/btToGreen"
            android:layout_marginLeft="70dp"
            android:layout_width="wrap_content"
            android:layout_height="wrap_content"
```

```xml
            android:text="变绿"
            android:onClick="doClick"/>
        <Button
            android:id="@+id/btToBlue"
            android:layout_marginLeft="70dp"
            android:layout_width="wrap_content"
            android:layout_height="wrap_content"
            android:text="变蓝"
            android:onClick="doClick"/>
    </LinearLayout>
</LinearLayout>
    <LinearLayout
        android:layout_width="fill_parent"
        android:layout_height="0dp"
        android:layout_weight="1.0">

    <TextView
        android:id="@+id/tvDisplay"
        android:layout_width="fill_parent"
        android:layout_height="fill_parent"
        android:gravity="center_horizontal|center_vertical"
        android:textSize="40dp"
        android:text="效果显示区"/>
    </LinearLayout>
    <LinearLayout
        android:layout_width="fill_parent"
        android:layout_height="0dp"
        android:layout_weight="1.0"
        android:orientation="vertical" >

    <LinearLayout
        android:layout_width="fill_parent"
        android:layout_height="70dp"
```

```xml
            android:layout_marginTop="30dp"
            android:gravity="center_horizontal">
        <Button
            android:id="@+id/btRegister"
            android:layout_width="wrap_content"
            android:layout_height="wrap_content"
            android:text="注册"
            android:onClick="doClick"/>
        <Button
            android:id="@+id/btCancel"
            android:layout_marginLeft="70dp"
            android:layout_width="wrap_content"
            android:layout_height="wrap_content"
            android:text="取消"
            android:onClick="doClick"/>
    </LinearLayout>
        <TextView
            android:layout_width="fill_parent"
            android:layout_height="wrap_content"
            android:gravity="center_horizontal"
            android:textSize="40dp"
            android:text="广播接收器"/>
    </LinearLayout>
</LinearLayout>
```

3. Java 文件的开发

（1）新建一个广播接收器类，该类在收到系统启动的广播后，将会启动本程序。

```java
public class MyReceiver extends BroadcastReceiver {
    @Override
    public void onReceive(Context context, Intent intent) {
        // TODO Auto-generated method stub
        Intent i = new Intent(context,MainActivity.class);
        i.setFlags(Intent.FLAG_ACTIVITY_NEW_TASK);
```

```
            String action = intent.getAction();
            if(Intent.ACTION_BOOT_COMPLETED.equals(action)){
                context.startActivity(i);
            }
        }
    }
}
```

(2)创建一个广播接收器内部类,收到不同的广播后,进行相应的操作。

```
    private class   InnerReceiver extends BroadcastReceiver{
        @Override
        public void onReceive(Context context, Intent intent) {
            // TODO Auto-generated method stub
            //获得action
            String action=intent.getAction();
            //根据不同的action 执行不同的操作
            if("com.xunfang.broadcast.tored".equals(action)){
                tvDisplay.setBackgroundColor(Color.RED);
            }else if("com.xunfang.broadcast.togreen".equals(action)){
                tvDisplay.setBackgroundColor(Color.GREEN);
            }else if("com.xunfang.broadcast.toblue".equals(action)){
                tvDisplay.setBackgroundColor(Color.BLUE);
            }

        }

    }
```

(3)初始化该广播接收器内部类。

```
    //自定义广播接收器内部类的初始化
    innerReceiver=new InnerReceiver();
    //意图过滤器的初始化
    intentFilter=new IntentFilter();
    intentFilter.addAction("com.xunfang.broadcast.tored");
    intentFilter.addAction("com.xunfang.broadcast.togreen");
    intentFilter.addAction("com.xunfang.broadcast.toblue");
```

```
//当前广播接收器是否已经注册的标识的初始化,初始值为 false
registerornot=false;
}
```

(4) 通过按钮动态注册广播接收器。

```
//为界面按钮添加事件处理方法
public void doClick(View v){
switch(v.getId()){
//首先判断当前接收器是否已经注册，没有注册则注册，否则提示已经注册
case R.id.btRegister:
    if(registerornot){
      Toast.makeText(this,"广播接收器已经注册，请勿重复注册",2000).show();
      return;
    }else{
    registerReceiver(innerReceiver, intentFilter);
    registerornot=true;
    Toast.makeText(this,"广播接收器注册成功",2000).show();
        }
    break;
}
```

(5) 通过按钮动态取消广播接收器。

```
case R.id.btCancel:
    if(registerornot){
        unregisterReceiver(innerReceiver);
        registerornot=false;
        Toast.makeText(this,"广播接收器被撤销",2000).show();
        return;
    }else{
        Toast.makeText(this,"广播接收器并未注册",2000).show();
    }
    break;
}
```

(6) 通过按钮发送广播。

```
public void doClick(View v){
```

项目十三 BroadCastReceiver 实现广播的接收与发送

```
switch(v.getId()){
//发送变红广播
case R.id.btToRed:
    Intent redintent=new Intent("com.xunfang.broadcast.tored");
    sendBroadcast(redintent);
    break;
//发送变绿广播
case R.id.btToGreen:
    Intent greenintent=new Intent("com.xunfang.broadcast.togreen");
    sendBroadcast(greenintent);
    break;
//发送变蓝广播
case R.id.btToBlue:
    Intent blueintent=new Intent("com.xunfang.broadcast.toblue");
    sendBroadcast(blueintent);
    break;
}
```

（7）程序保存后，运行测试，结果如图13-3和图13-4所示。

初始效果

单击【注册】按钮后

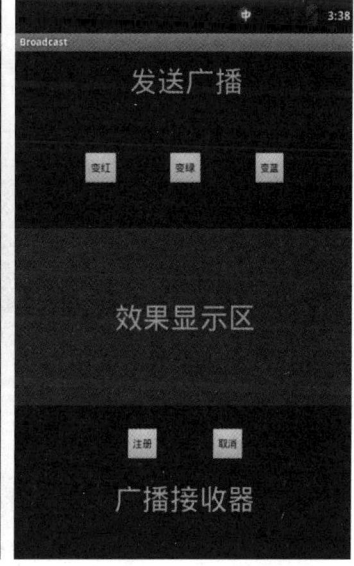
接着单击【红色】按钮

图13-3 运行结果

接着单击【绿色】按钮　　　　　　　单击【取消】按钮

图 13-4　运行结果

四、项目思考与扩展

1. 修改程序，在程序中加入一个文本输入框，在发送广播时把文本输入框中的内容发送出去并显示在界面里，如图 13-5 所示。

图 13-5　发送信息

2．创建一个广播接收者，并注册广播接收者为全局的，监听程序安装成功的广播（PACKAGE_ADDED_ACTION）接收到广播后在后台打印消息，编译运行程序，关闭程序，安装一个程序查看结果，如图13-6所示。

Application	Tag	Text
com.xunfang.androidcode09b	System.out	卸载了程序package:com.tencent.mm
com.xunfang.androidcode09b	System.out	安装了程序package:com.tencent.mm
com.xunfang.androidcode09b	System.out	卸载了程序package:com.taou.avatar
com.xunfang.androidcode09b	System.out	安装了程序package:com.taou.avatar
com.xunfang.androidcode09b	System.out	卸载了程序package:com.baidu.input
com.xunfang.androidcode09b	System.out	安装了程序package:com.baidu.input
com.xunfang.androidcode09b	System.out	卸载了程序package:com.UCMobile
com.xunfang.androidcode09b	System.out	安装了程序package:com.UCMobile

图13-6 监听程序安装与卸载

项目十四　Service 生命周期

【本章导读】

　　Service 是 Android 四大组件中与 Activity 最相似的组件，它们都代表可执行的程序。Service 一旦被启动起来，就与 Activity 一样，完全具有自己的生命周期。Service 与 Activity 的区别在于：Service 一直在后台运行，它没有用户界面，所以绝不会到前台来。因此，如果某个程序组件需要在运行时向用户呈现某种界面，或者该程序需要与用户交互，就需要使用 Activity，否则就应该考虑使用 Service 了。

一、项目要求

1. 了解 Service 是什么，有什么用途。
2. 掌握 Service 的生命周期，了解大体工作流程。
3. 熟悉 Service 的两种启动方式，学会使用 Service。

二、项目相关知识

1. Service 简介

　　Service 是 Android 系统中的四大组件之一，它是一种长生命周期的、没有可视化界面、运行于后台的一种服务程序。比如我们播放音乐的时候，有可能想边听音乐边干些其他事情，当退出播放音乐的应用，如果不用 Service，我们就听不到歌了。

　　服务主要用于两个目的：

➢ 后台运行和跨进程访问。通过启动一个服务，可以在不显示界面的前提下在后台运行指定的项目，这样可以不影响用户做其他的事情。
➢ 通过 AIDL 服务可以实现不同进程之间的通信，这也是服务的重要用途之一。

2. Service 生命周期

Service 的生命周期并不像 Activity 那么复杂，它只继承了 onCreate()，onStartCommend()，onDestroy()三个方法，当第一次启动 Service 时，先后调用了 onCreate()和 onStart()这两个方法，当停止 Service 时，则执行 onDestroy()方法，这里需要注意的是，如果 Service 已经启动了，当我们再次启动 Service 时，不会再执行 onCreate()方法，而是直接执行 onStart()方法。

Android 下的 Service 生命周期分为未绑定 Activity 的 service 和绑定了 Activity 的 service。两种不同 service 服务的生命周期图示如图 14-1 所示。

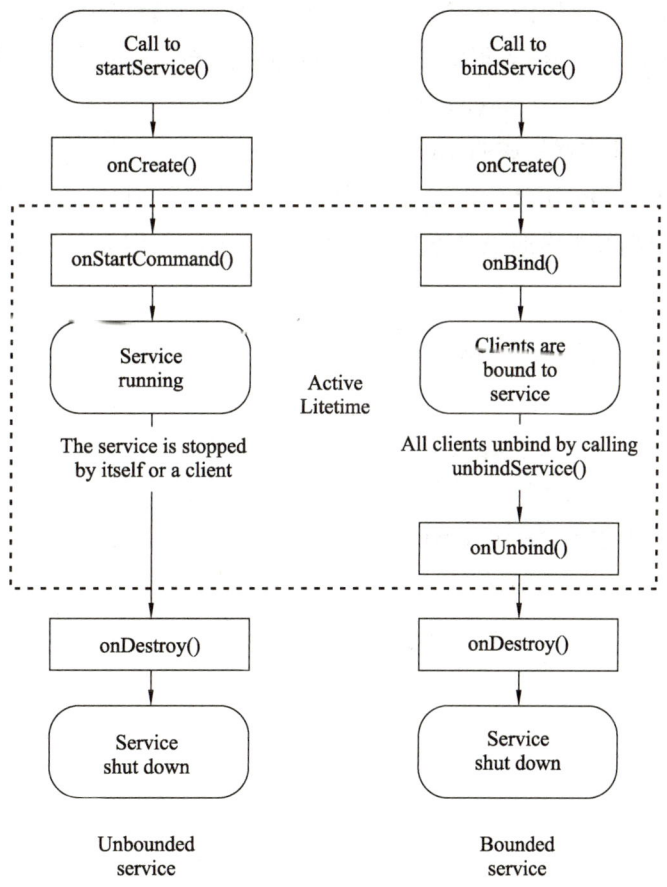

图 14-1　service 生命周期示意图

未绑定 Activity 的 Service 必须通过 startService()方法启动，生命周期依次包含了创建

onCreate()、开始 onStartCommand()、销毁 onDestory() 三个方法。其中创建方法 onCreate() 和销毁方法 onDestory() 只被调用一次，开始方法 onStartCommand() 方法可以被调用多次。

绑定了 Activity 的 Service 通过 bindService() 方法启动，生命周期包括创建 onCreate()、绑定 onBind()、解绑 onUnbind() 和销毁 onDestory() 四个方法。其中创建和销毁、绑定和解绑方法是对应的，都只执行一次。

3. 跨进程服务简介

什么是 AIDL？AIDL 全称是 Android Interface Definition Language，这是一种接口定义语言，采用远程过程调用（Remote Procedure Call，RPC）方式实现。这些服务可以被其他应用程序访问。

建立 AIDL 服务分为以下几个步骤：

（1）在 Android 工程的 Java 源目录中建立一个扩展名为 aidl 的文件。
（2）建立 aidl 文件的内容正确，ADT 会自动生成一个 Java 接口文件（*.Java）。
（3）建立一个服务类（Service 子类）。
（4）实现由 aidl 文件生成的 Java 接口。
（5）在 AndroidManifest.xml 文件中配置 AIDL 服务，<action> 标签中 android:name 的属性值就是客户端要引用的该服务的 ID，也就是 Intent 类构造方法的参数值。

三、项目实施过程

1. 工程创建

参考前面实验中创建"Helloworld"工程的步骤创建一个工程（如 AndroidCode14），工程目录结构，如图 14-2 所示。

图 14-2　工程目录结构图

2. XML 布局文件的开发

（1）界面采用相对布局管理器（垂直），里面再包含一个表格布局（水平），这个表格布局里包含四个 Button（分别是启动服务、停止服务、绑定服务、解绑服务），表格布局下面是一个 Button（退出程序），界面如图 14-3 所示。

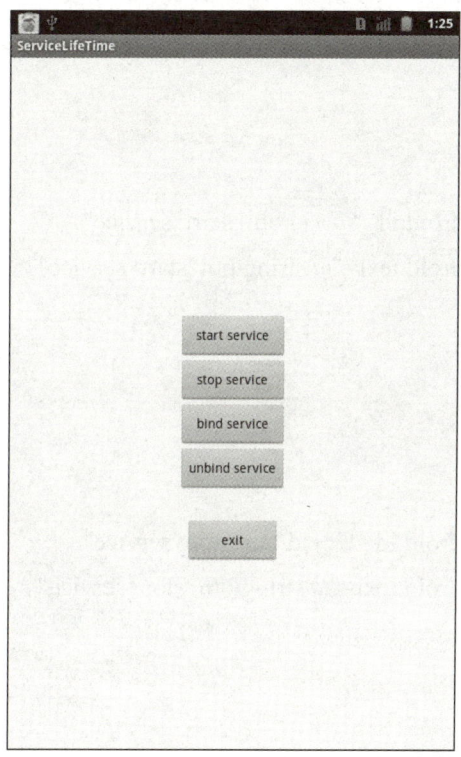

图 14-3　界面设置

（2）界面设计完整代码：

<RelativeLayout xmlns:android="http://schemas.android.com/apk/res/android"
　　xmlns:tools="http://schemas.android.com/tools"
　　android:layout_width="match_parent"
　　android:layout_height="match_parent"
　　android:paddingBottom="@dimen/activity_vertical_margin"
　　android:paddingLeft="@dimen/activity_horizontal_margin"
　　android:paddingRight="@dimen/activity_horizontal_margin"
　　android:paddingTop="@dimen/activity_vertical_margin"
　　tools:context=".MainActivity" >

```xml
<TableLayout
        android:id="@+id/layout_top"
        android:layout_width="wrap_content"
        android:layout_height="wrap_content"
        android:layout_centerHorizontal="true"
        android:layout_centerVertical="true" >
<!-- 开始 service -->
<TableRow>
        <Button
                android:id="@+id/btn_start_service"
                android:text="@string/btn_start_service" >
        </Button>
</TableRow>
<!-- 结束 service -->
<TableRow>
        <Button
                android:id="@+id/btn_stop_service"
                android:text="@string/btn_stop_service" />
</TableRow>
<!-- 绑定 service -->
<TableRow>
        <Button
                android:id="@+id/btn_bind_service"
                android:text="@string/btn_bind_service" />
</TableRow>
<!-- 解绑 service -->
<TableRow>
        <Button
                android:id="@+id/btn_unbind_service"
                android:text="@string/btn_unbind_service" />
</TableRow>
</TableLayout>
        <Button
```

```
            android:id="@+id/btn_exit"
            android:layout_width="wrap_content"
            android:layout_height="wrap_content"
            android:layout_below="@id/layout_top"
            android:layout_centerHorizontal="true"
            android:layout_margin="30dp"
            android:text="@string/btn_exit" />
</RelativeLayout>
```

（3）按【Ctrl+S】组合键保存后退出，至此界面文件就编写完成了。

3. Java 文件的开发

程序由三个类文件组成：MainActivity（主界面），MyBroadcastReceiver（广播类）和 MyService（服务类）。广播类用户接收程序发送的信息并处理相关操作。

（1）在 MainActivity 中定义开启服务、停止服务、绑定服务、解绑服务和退出 5 个 Button 按钮并初始化。初始化控件对象，使程序运行时候的控件显示和 XML 文件中绘制好的控件关联起来，对控件的操作就是对图形化界面上按钮的操作：

```
Button btnStartService, btnStopService, btnBindService, btnUnbindService,btnExit;
```

给 5 个 Button 按钮设定监听器，单击按钮时触发监听器，对应不同按钮监听器做出不同的响应。

```
btnStartService = (Button) findViewById(R.id.btn_start_service);
btnStopService = (Button) findViewById(R.id.btn_stop_service);
btnBindService = (Button) findViewById(R.id.btn_bind_service);
btnUnbindService = (Button) findViewById(R.id.btn_unbind_service);
btnExit=(Button)findViewById(R.id.btn_exit);
```

对应五个按钮的 intent 服务，分别对应启动服务、停止服务、绑定服务和解绑服务。

```
    public void onClick(View v) {
        Intent intent = new Intent();
        intent.setClass(context, MyService.class);
        switch (v.getId()) {
        // 启动服务
        case R.id.btn_start_service:
            context.startService(intent);
            break;
```

```
            // 停止服务
            case R.id.btn_stop_service:
                context.stopService(intent);
                break;
            // 绑定服务
            case R.id.btn_bind_service:
                context.bindService(intent, serviceConnection, BIND_AUTO_CREATE);
                break;
            // 解绑服务
            case R.id.btn_unbind_service:
                context.unbindService(serviceConnection);
                break;
            case R.id.btn_exit:
                unregisterReceiver(receiver);
                finish();
                break;
            default:
                break;
        }
    }
```

（2）在 MyBroadcastReceiver 类中重写 onReceive 方法，该方法用于接收 update 更新的广播信息，当程序收到内容为"update"的 action 时输出当前时间：

```
public void onReceive(Context context, Intent intent) {
    // TODO Auto-generated method stub
    String action = intent.getAction();
    if (action.equals("update")) {
        String time = intent.getExtras().getString("time");
        System.out.println(time);
        System.out.println("===>"+mContext);
    }
}
```

（3）在 MyService 类中的关键代码说明：Service 类里的代码就是单击不同 Button 按

钮时所触发的相应操作,此服务在后台运行。在 MyService 类中编写代码,实现开启服务,关闭服务,绑定服务,解绑服务的方法。新建一个 Thread 线程每隔一秒钟输出当前的精确时间,并用 intent 传递数据给接收器使用。

```java
public void onCreate() {
    Log.i("TAG", "启动服务");
    Toast.makeText(mContext, "启动服务", Toast.LENGTH_SHORT).show();
    // 开启计数
    new Thread(new Runnable() {
        @Override
        public void run() {
            // TODO Auto-generated method stub
            while (!flag) {
                try {
                    Thread.sleep(1000);
                    Log.i("TAG", "now time===>" + getSystemTime());
                    // 发送消息
                    String str = "current time:"+getSystemTime();
                    intent = new Intent();
                    intent.setAction("update");
                    intent.putExtra("time", str);
                    mContext.sendBroadcast(intent);

                } catch (Exception e) {
                    // TODO: handle exception
                }
            }
        }
    }).start();
    super.onCreate();
}
```

在 onDestory()方法中新建一个 Thread 线程暂停 1s,然后打印输出"服务停止":

```java
@Override
public void onDestroy() {
```

```
        flag = true;
        try {
            Thread.sleep(1000);
        } catch (InterruptedException e) {
            // TODO Auto-generated catch block
            e.printStackTrace();
        }
        Log.i("TAG", "停止服务");
        Toast.makeText(mContext, "停止服务", Toast.LENGTH_SHORT).show();
        super.onDestroy();
    }
```

在 onStartCommand 方法中输出开始的命令,在 log 输出结果中可以看到当程序结束时候,不管是否已经关闭服务,服务都会被终止。

```
public int onStartCommand(Intent intent, int flags, int startId) {
        Log.i("TAG", "onStartCommand()...");
        return super.onStartCommand(intent, flags, startId);
}
```

onUnbind 方法用于解除服务的绑定,当服务绑定解除,程序退出以后服务也跟着退出。
AndroidManifest 文件的配置:因为需要使用自定义的服务和全局广播,所以必须在 AndroidManifest 中添加注册服务和广播的声明:

```xml
<service
        android:name=".MyService"
        android:exported="true" >
</service>

<receiver android:name=".MyBroadcastReceiver" >
<intent-filter>
        <action android:name="update" />
</intent-filter>
</receiver>
```

4. 扩展练习

（1）在 service 中加入一个循环打印消息，通过 startService 方式启动服务，查看 LogCat 输出信息，关闭程序查看 LogCat 中的信息，如图 14-4 所示。

Level	Time	PID	TID	Application	Tag	Text
I	01-02 06:37:17.770	4095	4095	com.xunfang.servicelifetime	System.out	启动服务
I	01-02 06:37:18.800	4095	4095	com.xunfang.servicelifetime	System.out	current time:1970C
I	01-02 06:37:18.800	4095	4095	com.xunfang.servicelifetime	System.out	current time:1970C
I	01-02 06:37:19.810	4095	4095	com.xunfang.servicelifetime	System.out	current time:1970C
I	01-02 06:37:19.810	4095	4095	com.xunfang.servicelifetime	System.out	current time:1970C
I	01-02 06:37:20.820	4095	4095	com.xunfang.servicelifetime	System.out	current time:1970C
I	01-02 06:37:20.820	4095	4095	com.xunfang.servicelifetime	System.out	current time:1970C

图 14-4　循环输出 service 信息

（2）使用 bindService 方式启动服务，查看 LogCat 输出信息，关闭程序查看 LogCat 输出信息，如图 14-5 所示。

Level	Time	PID	TID	Application	Tag	Text
I	01-02 00:30:05.700	3301	3301	com.xunfang.servicelifetime	System.out	启动服务
I	01-02 00:30:05.720	3301	3301	com.xunfang.servicelifetime	System.out	绑定服务
I	01-02 00:30:06.730	3301	3301	com.xunfang.servicelifetime	System.out	current time:1970C
I	01-02 00:30:06.730	3301	3301	com.xunfang.servicelifetime	System.out	current time:1970C
I	01-02 00:30:07.740	3301	3301	com.xunfang.servicelifetime	System.out	current time:1970C
I	01-02 00:30:07.740	3301	3301	com.xunfang.servicelifetime	System.out	current time:1970C
I	01-02 00:30:09.060	3301	3301	com.xunfang.servicelifetime	System.out	解绑服务
I	01-02 00:30:10.060	3301	3301	com.xunfang.servicelifetime	System.out	停止服务

图 14-5　bindService 方式服务

（3）在程序中加入例如：log.d("debug","1")，按照 service 中各回调函数执行的顺序不同，依次在 LogCat 中打印输出 1、2、3、4，如图 14-6 所示。

Level	Time	PID	TID	Application	Tag	Text
D	01-02 00:34:38.650	3301	3301	com.xunfang.servicelifetime	TAG	启动服务-->1
D	01-02 00:34:38.660	3301	3301	com.xunfang.servicelifetime	TAG	绑定服务-->2
D	01-02 00:34:42.930	3301	3301	com.xunfang.servicelifetime	TAG	解绑服务-->3
D	01-02 00:34:43.930	3301	3301	com.xunfang.servicelifetime	TAG	停止服务-->4

图 14-6　生命周期执行顺序

5. 项目验证

（1）将 Android 开发终端与 PC 机相连。单击程序图标进入程序，单击“start service”按钮开启服务，并查看 log 输出顺序是否如图 14-7 所示。

Level	Time	PID	TID	Application	Tag	Text
I	01-02 00:36:26.890	3301	3301	com.xunfang.servicelifetime	System.out	启动服务
I	01-02 00:36:27.930	3301	3301	com.xunfang.servicelifetime	System.out	current time:1970C
I	01-02 00:36:27.930	3301	3301	com.xunfang.servicelifetime	System.out	current time:1970C
I	01-02 00:36:28.930	3301	3301	com.xunfang.servicelifetime	System.out	current time:1970C
I	01-02 00:36:28.930	3301	3301	com.xunfang.servicelifetime	System.out	current time:1970C
I	01-02 00:36:29.940	3301	3301	com.xunfang.servicelifetime	System.out	current time:1970C
I	01-02 00:36:29.940	3301	3301	com.xunfang.servicelifetime	System.out	current time:1970C
I	01-02 00:36:30.950	3301	3301	com.xunfang.servicelifetime	System.out	current time:1970C

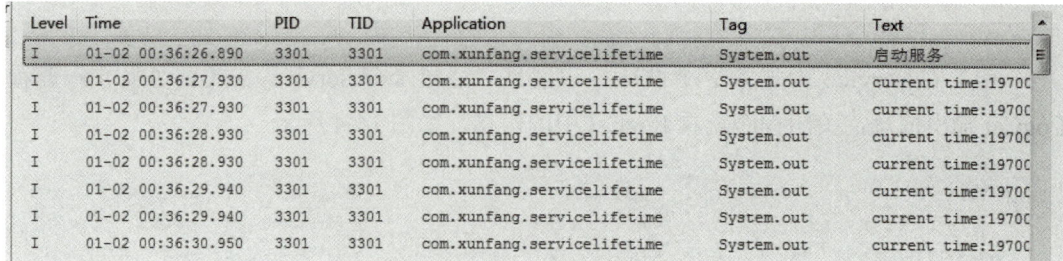

图 14-7　开启服务后的 log 显示

（2）单击“stop service”停止服务，查看 LogCat 输出是否如图 14-8 所示。

Level	Time	PID	TID	Application	Tag	Text
I	01-02 00:37:12.430	3301	3301	com.xunfang.servicelifetime	System.out	current time:1970C
I	01-02 00:37:12.430	3301	3301	com.xunfang.servicelifetime	System.out	current time:1970C
I	01-02 00:37:13.440	3301	3301	com.xunfang.servicelifetime	System.out	current time:1970C
I	01-02 00:37:13.440	3301	3301	com.xunfang.servicelifetime	System.out	current time:1970C
I	01-02 00:37:14.450	3301	3301	com.xunfang.servicelifetime	System.out	current time:1970C
I	01-02 00:37:14.450	3301	3301	com.xunfang.servicelifetime	System.out	current time:1970C
I	01-02 00:37:16.280	3301	3301	com.xunfang.servicelifetime	System.out	停止服务

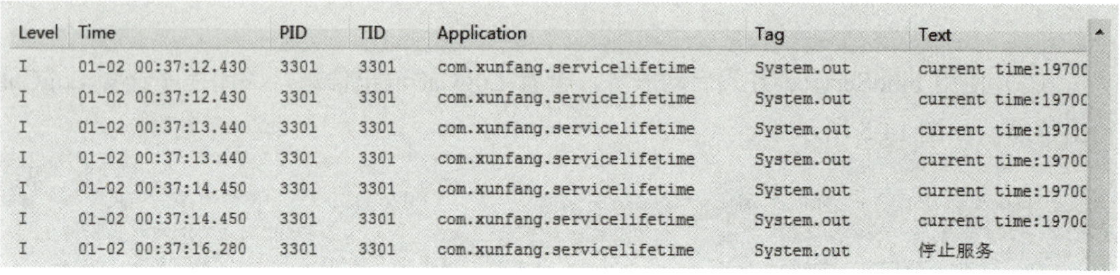

图 14-8　停止服务的 log 显示

（3）在开启了服务的情况下直接单击“e 运行。重新进入程序，单击“bind service”按钮绑定程序，查看 log 输出顺序，如图 14-9 所示。xit”退出程序看 log 日志是否 service 服务还在。

Level	Time	PID	TID	Application	Tag	Text
I	01-02 00:40:38.980	3572	3572	com.xunfang.servicelifetime	System.out	启动服务
I	01-02 00:40:39.010	3572	3572	com.xunfang.servicelifetime	System.out	绑定服务
I	01-02 00:40:39.020	3572	3572	com.xunfang.servicelifetime	TAG	服务连接成功!
I	01-02 00:40:40.020	3572	3572	com.xunfang.servicelifetime	System.out	current time:1970C
I	01-02 00:40:40.020	3572	3572	com.xunfang.servicelifetime	System.out	current time:1970C
I	01-02 00:40:41.030	3572	3572	com.xunfang.servicelifetime	System.out	current time:1970C
I	01-02 00:40:41.030	3572	3572	com.xunfang.servicelifetime	System.out	current time:1970C
I	01-02 00:40:42.040	3572	3572	com.xunfang.servicelifetime	System.out	current time:1970C

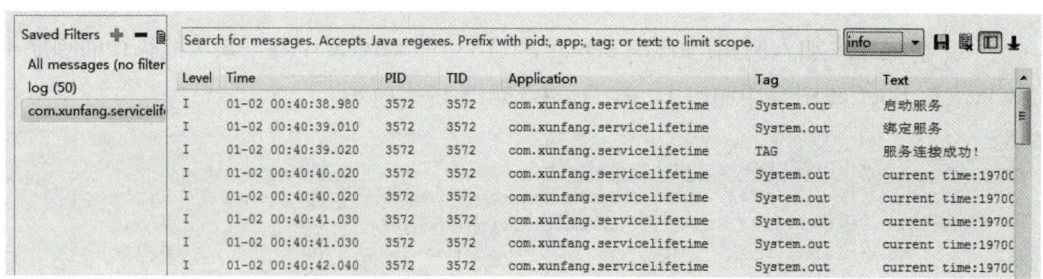

图 14-9　绑定服务后的 log 显示

（4）单击“unbind service”按钮解绑服务，查看 log 信息输出是否如图 14-10 所示。

Level	Time	PID	TID	Application	Tag	Text
I	01-02 00:41:33.630	3572	3572	com.xunfang.servicelifetime	System.out	current time:19700
I	01-02 00:41:34.640	3572	3572	com.xunfang.servicelifetime	System.out	current time:19700
I	01-02 00:41:34.640	3572	3572	com.xunfang.servicelifetime	System.out	current time:19700
I	01-02 00:41:35.670	3572	3572	com.xunfang.servicelifetime	System.out	解绑服务
I	01-02 00:41:36.670	3572	3572	com.xunfang.servicelifetime	System.out	停止服务

图 14-10 解绑后的 log 显示

（5）在单击【bind service】按钮后直接单击【exit】退出程序，观察 log 日志查看 service 服务是否还在运行。对比开启服务和绑定服务后退出程序，思考有何异同。

四、项目思考与扩展

1．什么是 Service？
2．Service 生命周期有哪几种？
3．Service 生命周期中回调函数的执行过程？

数据处理篇

项目十五 SharedPerference 与 XML

【本章导读】

　　所有应用程序都必然涉及数据的输入、输出，Android 应用也不例外，应用程序的参数设置、程序运行状态数据这些都需要保存到外部存储器上，这样系统关机之后数据才不会丢失。Android 应用开发是使用 Java 语言来开发的，因此开发者在 Java IO 中的编程经验大部分都可"移植"到 Android 应用开发上，但 Android 系统还提供了一些专门的 IO API，通过这些 API 可以更有效地进行输入、输出。

　　例如有些时候，应用程序只有少量的数据需要保存，而且这些数据的格式很简单，都是普通的字符串、标量类型的值（比如是否打开音效、是否使用振动效果等），对于这种数据，Android 提供了 SharedPreferences 进行保存。

　　SAX（Simple API for XML）是指一种接口，或者一个软件包。SAX 事件驱动型的 XML 解析方式，顺序读取 XML 文件，不需要一次全部装载整个文件。当遇到像文件开头、文档结束或者标签开头与标签结束时，会触发一个事件，用户通过在其回调事件中写入处理代码来处理 XML 文件，适合对 XML 的顺序访问，且是只读的。由于移动设备的内存资源有限，SAX 的顺序读取方式更适合移动开发。

一、项目要求

1. 掌握使用 SharedPreferences 存储和获取数据的方法。
2. 掌握 SAX 创建和解析 XML 文件的方法。
3. 了解 SharedPreferences 实现原理。

项目十五　SharedPerference 与 XML

二、项目相关知识

1. SharedPerference 简介

SharedPreference（用户偏好设置）可以将数据保存在应用软件的私有存储区，这些存储区中的数据只能被写入这些数据的软件读取。实际上 sharedPreference 处理的就是一个 key-value 对。

获取 SharedPerference 对象有两种方法：一个是通过 Activity 对象获取，方法为 getPreferences（int mode）；另一种是通过 Context 对象获取，方法为 getSharedPreferences（String name,int mode）。

2. SAX 的简介

SAX 是一个解析速度快并且占用内存少的 XML 解析器，非常适合用于 Android 等移动设备。SAX 解析 XML 文件采用的是事件驱动，也就是说，它并不需要解析整个文档，在按内容顺序解析文档的过程中，SAX 会判断当前读到的字符是否为合法 XML 语法中的某部分，如果符合就会触发事件。所谓事件，其实就是一些回调（callback）方法，这些方法（事件）定义在 ContentHandler 接口中。下面是一些 ContentHandler 接口常用的方法：

➢ startDocument()

当遇到文档开头的时候，调用这个方法，可以在其中做一些预处理的工作。

➢ endDocument()

和上面的方法相对应，当文档结束的时候，调用这个方法，可以在其中做一些善后的工作。

➢ startElement(String namespaceURI,String localName,String qName,Attributes atts)

当读到一个开始标签的时候，会触发这个方法。namespaceURI 就是命名空间，localName 是不带命名空间前缀的标签名，qName 是带命名空间前缀的标签名。通过 atts 可以得到所有的属性名和相应的值。

要注意的是 SAX 中一个重要的特点就是它的流式处理，当遇到一个标签的时候，它并不会记录下以前所碰到的标签，也就是说，在 startElement()方法中，所有你所知道的信息，就是标签的名字和属性，至于标签的嵌套结构，上层标签的名字，是否有子元素等等其他与结构相关的信息，都是不得而知的，都需要你的程序来完成。

➢ endElement(String uri,String localName,String name)

这个方法和上面的方法相对应，在遇到结束标签的时候，调用这个方法。

➢ characters(char[]ch,int start,int length)

这个方法用来处理在 XML 文件中读到的内容，第一个参数为文件的字符串内容，后面两个参数是读到的字符串在这个数组中的起始位置和长度，使用 new String(ch,start,length)就可以获取内容。

三、项目实施过程

1. 创建工程

在 eclipse 中新建一个名为 AndroidCode15 的 Android 项目工程，在工程下新建名为 com.xdxy.sharedpreferecedemo 的包，在包下新建 MainActivity、Product 和 XML2Product 三个 Java 文件，如图 15-1 所示。

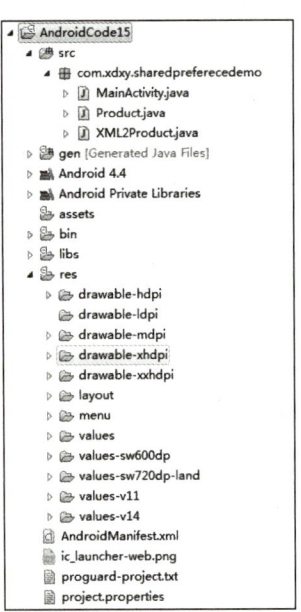

图 15-1　工程目录结构图

2. XML 布局文件的开发

界面布局如图 15-2 所示，总体布局采用 RelativeLayout，其内部再分为第 1 区（通过

Activity 方式调用 sharedPerference 操作),第 2 区(通过 Context 方式调用 sharedPerference 操作),第 3 区(读取 XML 文件),第 4 区(退出按钮)四个部分。第 1 区部分采用 RelativeLayout 布局,包括一个 Textview 显示标题,一个 TableLayout 内含两个 Button 按钮(写入和读取)。第 2 区部分和第 1 区一样,第 3、第 4 区部分都只有一个 Button 按钮。

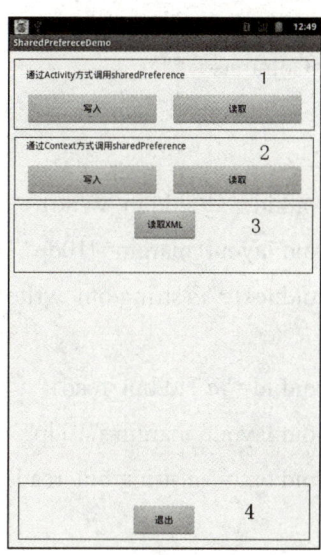

图 15-2　界面设置

下面我们开始编写实现布局框架的代码。

(1)第 1 区部分的布局是在 RelativeLayout 布局中放置一个 TextView 和 TableLayout(包含两个 Button 控件),代码如下:

```
<RelativeLayout
        android:id="@+id/layout_top"
        android:layout_width="wrap_content"
        android:layout_height="wrap_content"
        android:layout_centerHorizontal="true" >
<!-- 方法一 -->
        <TextView
            android:id="@+id/tv_title1"
            android:layout_width="wrap_content"
            android:layout_height="wrap_content"
            android:layout_margin="10dp"
            android:text="@string/str_title1" />
```

```xml
<!-- 读取和写入按钮 -->
    <TableLayout
        android:layout_width="wrap_content"
        android:layout_height="wrap_content"
        android:layout_below="@id/tv_title1"
        android:stretchColumns="*" >
<TableRow tools:ignore="UselessParent" >
    <Button
            android:id="@+id/btn_write1"
            android:layout_margin="10dp"
            android:text="@string/btn_write1" />
    <Button
            android:id="@+id/btn_read1"
            android:layout_margin="10dp"
            android:text="@string/btn_read1" />
</TableRow>
</TableLayout>
</RelativeLayout>
```

（2）第 2 区部分布局和第 1 区部分一样，只需要修改 id 和显示文字部分即可，代码如下：

```xml
<RelativeLayout
        android:id="@+id/layout_mid"
        android:layout_width="wrap_content"
        android:layout_height="wrap_content"
        android:layout_below="@id/layout_top"
        android:layout_centerHorizontal="true" >
<TextView
            android:id="@+id/tv_title2"
            android:layout_width="wrap_content"
            android:layout_height="wrap_content"
            android:layout_margin="10dp"
            android:text="@string/str_title2" />
<!-- 读取和写入按钮 -->
```

```xml
<TableLayout
        android:layout_width="wrap_content"
        android:layout_height="wrap_content"
        android:layout_below="@id/tv_title2"
        android:stretchColumns="*" >
<TableRow tools:ignore="UselessParent" >
    <Button
                android:id="@+id/btn_write2"
                android:layout_margin="10dp"
                android:text="@string/btn_write1" />
    <Button
                android:id="@+id/btn_read2"
                android:layout_margin="10dp"
                android:text="@string/btn_read1" />
</TableRow>
</TableLayout>
</RelativeLayout>
```

（3）第3区部分是读取XML文件按钮，由一个Button按钮组成，代码如下：

```xml
<!-- 显示XML的按钮 -->
<Button
        android:layout_width="100dp"
        android:layout_height="wrap_content"
        android:layout_below="@id/layout_mid"
        android:layout_centerHorizontal="true"
        android:layout_margin="10dp"
        android:id="@+id/btn_show_xml"
        android:text="@string/btn_show_xml" />
```

（4）第4区部分是退出按钮，由一个Button控件组成，代码如下：

```xml
<!-- 退出按钮 -->
<Button
        android:id="@+id/btn_exit"
        android:layout_width="100dp"
        android:layout_height="wrap_content"
```

```
android:layout_alignParentBottom="true"
android:layout_centerHorizontal="true"
android:text="@string/btn_exit" />
```

（5）按【Ctrl+S】组合键保存后退出，至此界面文件就编写完成了。

3. Java 程序的开发

MainActivity 是主程序，作用是控制主界面上各个控件的作用。onCreate()方法在程序创建时候调用，initView()方法用于初始化控件，建立与 XML 文件中控件的对应关系。下面对其关键代码进行分析。

（1）以下代码定义文本显示控件和按钮控件。

```
// 控件
TextView tvTile1, tvTile2;
Button btnWrite1, btnRead1, btnWrite2, btnRead2, btnExit, btnShowXML;
```

（2）给控件设置监听器，使得当单击界面中不同控件时触发对应响应事件。

```
private void initListener() {
    btnWrite1.setOnClickListener(this);
    btnRead1.setOnClickListener(this);
    btnWrite2.setOnClickListener(this);
    btnRead2.setOnClickListener(this);
    btnExit.setOnClickListener(this);
    btnShowXML.setOnClickListener(this);
}
private void initView() {
    tvTile1 = (TextView) findViewById(R.id.tv_title1);
    btnWrite1 = (Button) findViewById(R.id.btn_write1);
    btnRead1 = (Button) findViewById(R.id.btn_read1);
    tvTile2 = (TextView) findViewById(R.id.tv_title2);
    btnWrite2 = (Button) findViewById(R.id.btn_write2);
    btnRead2 = (Button) findViewById(R.id.btn_read2);
    btnExit = (Button) findViewById(R.id.btn_exit);
    btnShowXML = (Button) findViewById(R.id.btn_show_xml);
}
```

（3）onClick()方法里包括对应界面中各个控件的操作处理方法，以下代码调用 Product

类的属性来显示产品信息。
```java
public void onClick(View v) {
    // TODO Auto-generated method stub
    switch (v.getId()) {
    case R.id.btn_show_xml:
        // 单击解析 xml 文件
        InputStream is = getResources().openRawResource(R.raw.products);
        XML2Product xml2Product = new XML2Product();
        try {
            // 转换成 utf-8 编码
            android.util.Xml.parse(is, Xml.Encoding.UTF_8, xml2Product);
        } catch (IOException e1) {
            // TODO Auto-generated catch block
            e1.printStackTrace();
        } catch (SAXException e1) {
            // TODO Auto-generated catch block
            e1.printStackTrace();
        }
        List<Product> products = xml2Product.getProducts();
        String msg = "共" + products.size() + "个产品\n";
        for (Product product : products) {
            msg += "id:" + product.getId() + "\t 产品名：" + product.getName()
                    + "\t 价格：" + product.getPrice() + "\n";
        }
        new AlertDialog.Builder(this).setTitle("产品信息").setMessage(msg)
                .setPositiveButton("关闭", null).show();
        break;
    }
}
```
（4）通过 Activity 方式调用 sharedPreference 的 Button 方法。
```java
case R.id.btn_write1:
    // 通过 Activity 的方式调用 sharedPreference
    // key-value 对的文件名 ,MODE_PRIVATE 默认方式能被所有 app 读取
    try {
```

```
            sharedPreferences = getPreferences(MODE_PRIVATE);
            // 加入 K-V 对
            editor = sharedPreferences.edit();
            editor.putString("name", "讯方通信技术有限公司");
            editor.commit();
            Toast.makeText(this, "写入成功!", Toast.LENGTH_SHORT).show();
        } catch (Exception e) {
            // TODO Auto-generated catch block
            Toast.makeText(this, "写入失败!", Toast.LENGTH_SHORT).show();
            e.printStackTrace();
        }
    break;
```

（5）通过 Context 方式调用 sharedPreference 写入信息。

```
    case R.id.btn_write2:
        // 通过 Context 的方式调用 sharedPreference
        // key-value 对的文件名 ,Context.MODE_PRIVATE 默认方式能被所有 app 读取
        try {
            sharedPreferences = getSharedPreferences("File1",
                    Context.MODE_PRIVATE);
            editor = sharedPreferences.edit();
            editor.putString("name", "Android 开发终端");
            editor.commit();
            Toast.makeText(this, "写入成功!", Toast.LENGTH_SHORT).show();
        } catch (Exception e) {
            // TODO Auto-generated catch block
            Toast.makeText(this, "写入失败!", Toast.LENGTH_SHORT).show();
            e.printStackTrace();
        }
    break;
```

Product 类是一个 JavaBean，存储产品类的对应属性（编号、产品名、售价），外部类通过 get 和 set 接口访问 Product 类的属性。

```
public class Product {
    private String id, name, price;
```

```java
    public String getId() {
        return id;
    }

    public void setId(String id) {
        this.id = id;
    }

    public String getName() {
        return name;
    }

    public void setName(String name) {
        this.name = name;
    }

    public String getPrice() {
        return price;
    }

    public void setPrice(String price) {
        this.price = price;
    }
}
```

XML2Product 类是对 xml 具体操作的方法,下面对关键代码进行说明。

(1) startDocument()方法用于开始分析 XML 文件。

```java
// 开始分析 xml
    @Override
    public void startDocument() throws SAXException {
        products = new ArrayList<Product>();
    }
```

(2) startElement()方法用于开始处理 xml 标签。

```
// 开始处理 xml 标签
```

```java
@Override
public void startElement(String uri, String localName, String qName,Attributes attributes)
throws SAXException {
    if (localName.equals("product")) {
        product = new Product();
    }
    super.startElement(uri, localName, qName, attributes);

}
```

（3）endElement()方法用于处理完标签的操作。

```java
// 处理完一个 xml 标签
@Override
public void endElement(String uri, String localName, String qName)
throws SAXException {
    if (localName.equals("product")) {
        products.add(product);
    } else if (localName.equals("id")) {
        product.setId(buffer.toString().trim());
        buffer.setLength(0);
    } else if (localName.equals("name")) {
        product.setName(buffer.toString().trim());
        buffer.setLength(0);
    } else if (localName.equals("price")) {
        product.setPrice(buffer.toString().trim());
        buffer.setLength(0);
    }
    super.endElement(uri, localName, qName);
}
```

（4）endDocument()方法用于处理完 xml 文件后的操作。

```java
// 处理完 xml 文件
@Override
public void endDocument() throws SAXException {
    // TODO Auto-generated method stub
```

项目十五 SharedPerference 与 XML

```
    super.endDocument();
}
```

（5）characters()方法用于读取字符分析点。

```
// 读取字符分析点
@Override
public void characters(char[] ch, int start, int length)throws SAXException {
    buffer.append(ch, start, length);
    super.characters(ch, start, length);
}
```

4. 扩展练习

（1）手动创建一个 XML 文件，参照程序给出的 XML 文件输入相应内容，然后使用程序的 XML 读取程序，读出 XML 文件的内容，如图 15-3 所示。

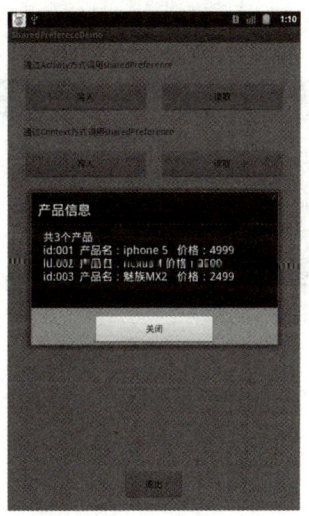

图 15-3　产品信息

（2）在 Eclipse 中使用 FileExplorer 中的 data/data/AndroidCode15 找到 SharedPreferences 的存储 XML 文件，导出查看内容，修改内容导入覆盖，用实验程序读取出来，如图 15-4～图 15-7 所示。

图 15-4　xml 文件保存位置

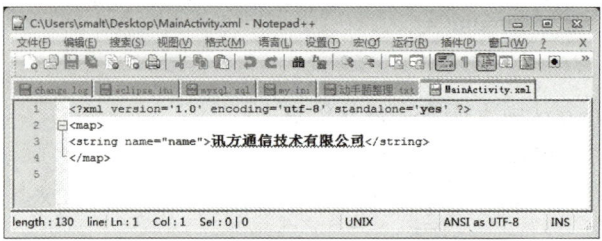

图 15-5　原始 xml 文件内容

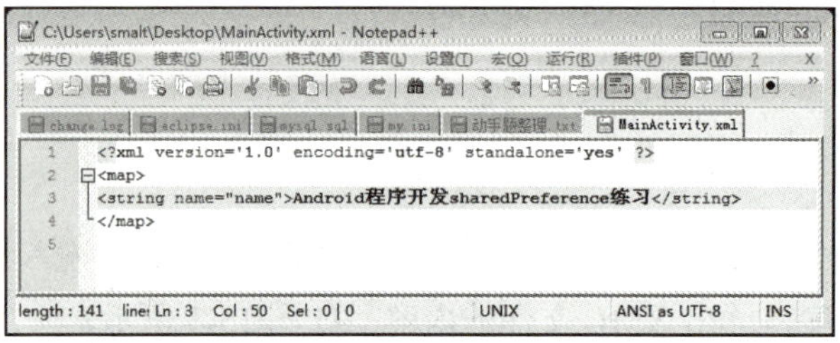

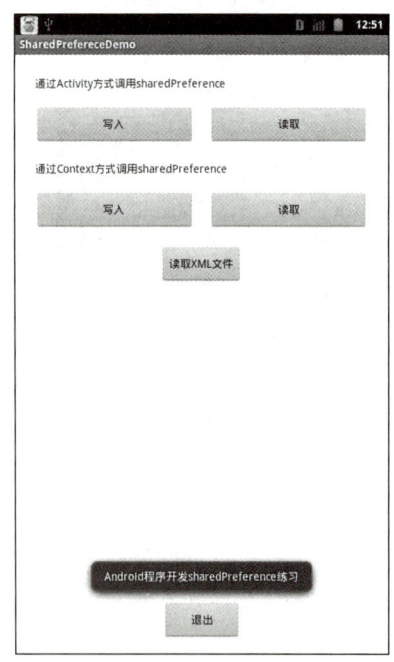

图 15-6　修改后的 xml 文件内容

图 15-7　运行程序读取 xml 文件的结果

5. 项目验证

（1）将 Android 开发终端与 PC 机相连。单击程序图标进入程序，在"通过 Activity 方式调用 sharedPerference"栏下单击【写入】按钮写入 sharedPerference，查看是否写入成功，如图 15-8 所示，然后单击【读取】按钮读取写入的信息，如图 15-9 所示。

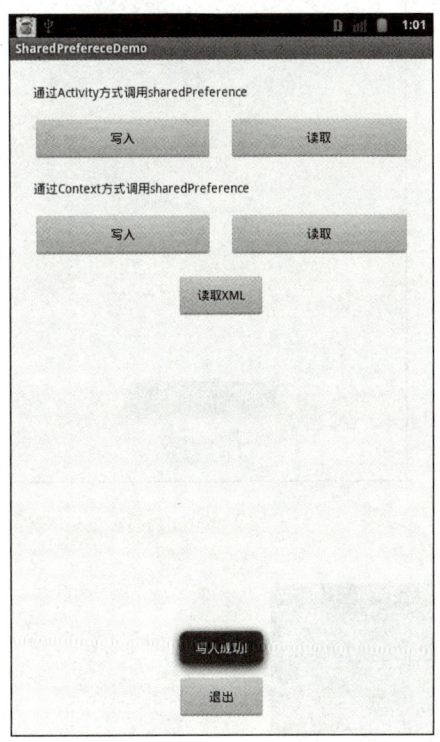
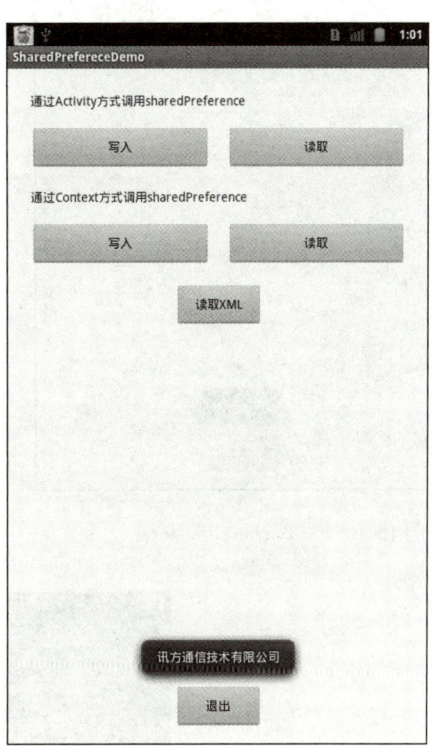

图 15-8　写入 sharedPerference　　　　　图 15-9　写入 sharedPerference

（2）单击程序图标进入程序，在"通过 Activity 方式调用 sharedPerference"栏下单击【写入】按钮写入 sharedPerference，查看是否写入成功，如图 15-10，然后单击【读取】按钮读取写入的信息，如图 15-11 所示。

（3）单击【读取 xml 文件】显示结果 SAX 引擎分析的 XML 文件内容，如图 15-12 所示。

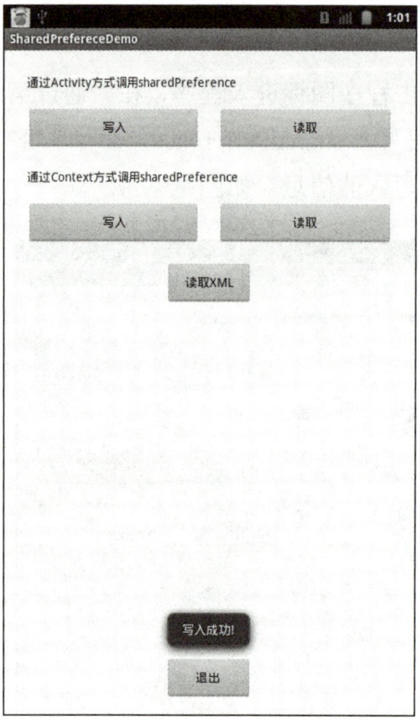

图 15-10　写入 sharedPerference 文件 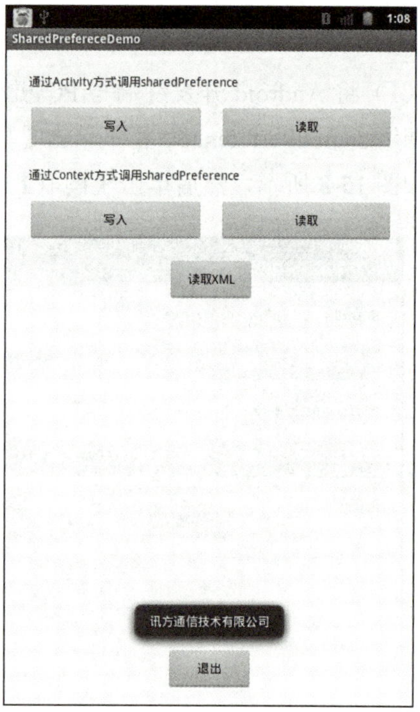　　　　图 15-11　读取 sharedPerference 文件

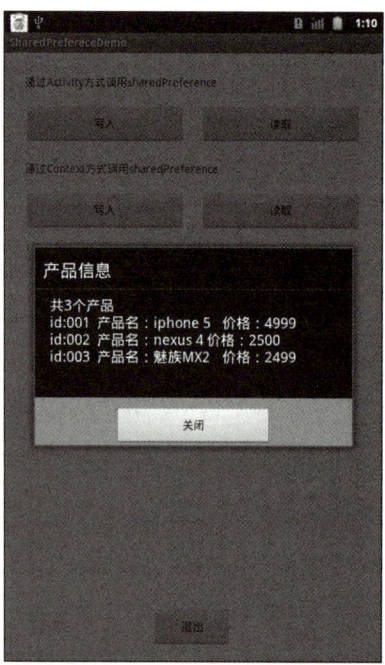

图 15-12　读取 XML 文件

四、项目思考与扩展

1. 什么是 SharedPerference？
2. 如何使用 SAX 引擎解析 XML 文件？
3. 如何使用字符流方式读取数据？

项目十六　IO 操作与数据存储访问

【本章导读】

读者学习 Java SE 的时候都知道 Java 提供了一套完整的 I/O 流体系，包括 FileInputStream、FileOutputStream 等，通过这些 I/O 流可以非常方便地访问磁盘上的文件内容。Android 同样支持以这种方式来访问手机存储器上的文件。

一、项目要求

1．通过 IO 操作实现对 Android 内部文件、外部存储设备中的文件以及资源文件的存储与访问。

2．掌握 Android 中的 IO 操作与文件操作模式。

3．熟悉 Android 设备操作权限声明。

4．了解资源文件的使用。

二、项目相关知识

1．内部文件存储

文件存储的核心就是输入和输出流，如果想对文件随心所欲地操作，直接使用流是最好的选择。读写采用 openFileOutput()和 openFileInput()方法。

openFileOutput()方法用如下形式返回一个 OutputStream 对象，并保存在 android 内存中的/data/data/包名/files 目录中，如图 16-1 所示。

```
□ □ com.xunfang.io                                    2000-01-01  07:52  drwxr-x--x
    □ □ files                                         2000-01-01  07:52  drwxrwx--x
        □ file.txt                               9    2000-01-01  07:52  -rw-rw----
```

图 16-1 内存文件存储

2. 外部文件存储

文件也可以存储在 SD 卡上，下面学习将图像文件保存在 SD 卡上并读取出来。使用 FileInputStream()和 FileOutputStream()来读写指定路径的文件。

Context 提供了如下两个方法来打开应用程序的数据文件夹里的文件 I/O 流。

- FileInputStream openFileInput(String name)：打开应用程序的数据文件夹下的 name 文件对应输入流。
- FileOutputStream openFileOutput(String name, int mode)：打开应用程序的数据文件夹下的 name 文件对应输出流。

上面两个方法分别用于打开文件输入流、输出流，其中第二个方法的第二个参数指定打开文件的模式，该模式支持如下值：

- MODE_PRIVATE：该文件只能被当前程序读写。
- MODE_APPEND：以追加方式打开该文件，应用程序可以向该文件中追加内容。
- MODE_WORLD_READABLE：该文件的内容可以被其他程序读取。
- MODE_WORLD_WRITEABLE：该文件的内容可由其他程序读、写。

除此之外，Context 还提供了如下几个方法来访问应用程序的数据文件夹：

- getDir(String name, int mode)：在应用程序的数据文件夹下获取或创建 name 对应的子目录。
- File getFilesDir()：获取该应用程序的数据文件夹的绝对路径。
- String[] fileList()：返回该应用程序的数据文件夹下的全部文件。
- deleteFile(String)：删除该应用程序的数据文件夹下的指定文件。

3. 资源文件使用

Android 程序使用的资源文件（文本、音频、图像）可以保存在 res/raw 或 res/assets 文件夹下。保持在这两个文件夹下的文件会原封不动地保存在 apk 包中，不会编译成二进制文件。

两个文件夹的区别如下：

- res/raw 中的文件会被映射到 R.Java 文件中，访问的时候直接使用资源 ID 即 R.id.filename；assets 文件夹下的文件不会被映射到 R.Java 中，访问的时候需要

AssetManager 类。

➢ res/raw 不可以有目录结构，而 assets 则可以有目录结构，也就是 assets 目录下可以再建立文件夹。

三、项目实施过程

下面我们通过一个具体例子来了解 I/O 操作和数据存储访问。

1. 创建工程

在 eclipse 中新建一个名为 AndroidCode16 的 Android 项目，新建名为 com.xdxy.io 的包，在包下新建 ExeActivity 和 MainActivity 两个 Java 文件，如图 16-2 所示。

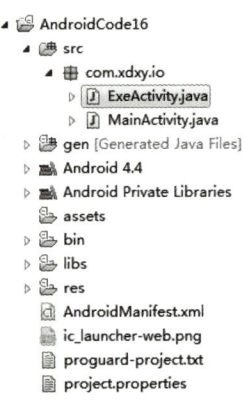

图 16-2　工程目录结构图

2. XML 布局文件的开发

界面布局如图 16-3 所示，总体布局采用 RelativeLayout，内布局为第 1 区，第 2 区，第 3 区，第 4 区四个部分。

第 1 区布局采用 RelativeLayout 布局，包括一个 Textview 显示标题，一个 TableLayout 内含两个 Button 按钮（写入和读取）；第 2 区布局和第 1 区布局一样；第 3 区布局是一个 ImageView 控件，用来显示 SD 卡中的图片；第 4 区布局是一个 Button 按钮，单击此按钮退出程序。

整体详细布局见源程序中的 activity_main.xml 文件，其中：

项目十六 IO 操作与数据存储访问

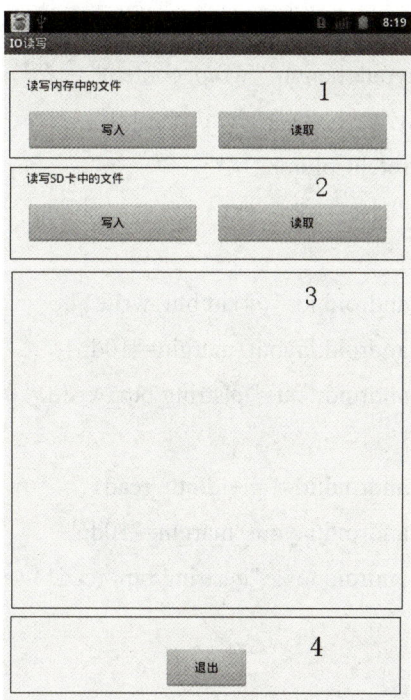

图 16-3 界面设置

（1）以下代码显示编号为第 1 区的布局，包括一个文本框 TextView 和两个按钮控件 Button。

```
<RelativeLayout
        android:id="@+id/layout_top"
        android:layout_width="wrap_content"
        android:layout_height="wrap_content"
        android:layout_centerHorizontal="true" >
<!-- 读写内存卡中数据 -->
<TextView
        android:id="@+id/tv_title1"
        android:layout_width="wrap_content"
        android:layout_height="wrap_content"
        android:layout_margin="10dp"
        android:text="@string/str_title1" />
<!-- 读取和写入按钮 -->
<TableLayout
```

```xml
            android:layout_width="wrap_content"
            android:layout_height="wrap_content"
            android:layout_below="@id/tv_title1"
            android:stretchColumns="*" >
    <TableRow tools:ignore="UselessParent" >
            <Button
                    android:id="@+id/btn_write1"
                    android:layout_margin="10dp"
                    android:text="@string/btn_write1" />
            <Button
                    android:id="@+id/btn_read1"
                    android:layout_margin="10dp"
                    android:text="@string/btn_read1" />
    </TableRow>
</TableLayout>
</RelativeLayout>
```

（2）以下代码显示编号为第 2 区的布局部分，由两个按钮控件 Button 组成。

```xml
        <!-- 读取和写入按钮 -->
<TableLayout
            android:id="@+id/tb_layout_sd"
            android:layout_width="wrap_content"
            android:layout_height="wrap_content"
            android:layout_below="@id/tv_title2"
            android:stretchColumns="*" >
    <TableRow tools:ignore="UselessParent" >
            <Button
                    android:id="@+id/btn_write2"
                    android:layout_margin="10dp"
                    android:text="@string/btn_write1" />
            <Button
                    android:id="@+id/btn_read2"
                    android:layout_margin="10dp"
                    android:text="@string/btn_read1" />
```

```
</TableRow>
</TableLayout>
```

（3）以下代码显示编号为第 3 区的布局，由一个图片显示控件 imageView 图像视图组成。

```
<ImageView
        android:id="@+id/iv_sd"
        android:layout_width="fill_parent"
        android:layout_height="fill_parent"
        android:layout_above="@id/btn_exit"
        android:layout_below="@id/tb_layout_sd"
        android:layout_centerHorizontal="true"
        android:layout_marginTop="10dp"
        android:contentDescription="@id/iv_sd" />
```

（4）以下代码显示编号为第 4 区的布局，由一个 Button 按钮组成。

```
<!-- 退出按钮 -->
<Button
    android:id="@+id/btn_exit"
    android:layout_width="100dp"
    android:layout_height="wrap_content"
    android:layout_alignParentBottom="true"
    android:layout_centerHorizontal="true"
    android:text="@string/btn_exit"/>
```

3. Java 文件的开发

程序比较简单，只有一个 MainActivity 文件，关键代码说明如下：

（1）initView()方法用于初始化控件，绑定 xml 中定义的 TextView、Button 和 ImageView 等。

```
// 控件
ImageView ivSD;
TextView tvTile1, tvTile2;
Button btnWrite1, btnRead1, btnWrite2, btnRead2, btnExit;
EditText etInput;
Context context;
```

```java
private void initView() {
    context = MainActivity.this;
    tvTile1 = (TextView) findViewById(R.id.tv_title1);
    btnWrite1 = (Button) findViewById(R.id.btn_write1);
    btnRead1 = (Button) findViewById(R.id.btn_read1);

    tvTile2 = (TextView) findViewById(R.id.tv_title2);
    btnWrite2 = (Button) findViewById(R.id.btn_write2);
    btnRead2 = (Button) findViewById(R.id.btn_read2);
    btnExit = (Button) findViewById(R.id.btn_exit);
    ivSD = (ImageView) findViewById(R.id.iv_sd);
}
```

（2）initListener()方法用于注册监听器，绑定控件，这样单击不同控件的时候会触发不同响应。

```java
private void initListener() {
    btnWrite1.setOnClickListener(this);
    btnRead1.setOnClickListener(this);

    btnWrite2.setOnClickListener(this);
    btnRead2.setOnClickListener(this);
    btnExit.setOnClickListener(this);
}
```

（3）write2Memeray()方法中使用输出字符流 OutputStream 的方式写入字符串，并在屏幕上给予提示。

```java
/**
 * @description 把文件写入内存
 * @param path
 *            文件地址
 */
private void write2Memery(String path) {
    try {
        OutputStream os = openFileOutput(MEM_PATH, Activity.MODE_PRIVATE);
```

```java
            // String str = "此条信息是从内存中读取。。";
            os.write(path.getBytes("utf-8"));
            os.close();
            Toast.makeText(context, "写入成功", Toast.LENGTH_SHORT).show();
        } catch (Exception e) {
            Toast.makeText(context, "读取失败！", Toast.LENGTH_SHORT).show();
        }
    }
```

（4）ReadFromMemery()方法通过输入字符流 InputStream 和字节流 buffer 方式从文件中读取字符串信息。

```java
    private void readFromMemery() {
        try {
            InputStream is = openFileInput(MEM_PATH);
            byte[] buffer = new byte[100];
            int byteCount = is.read(buffer);
            String str = new String(buffer, 0, byteCount, "utf-8");
            Toast.makeText(context, str, Toast.LENGTH_SHORT).show();
        } catch (Exception e) {
            Toast.makeText(context, "读取失败！", Toast.LENGTH_SHORT).show();
        }
    }
```

（5）write2SDCard()使用 FileOutputStream 方式把 jpg 格式的图片存入 SD 卡指定的路径。

```java
    private void write2SDCard() {
        try {
            FileOutputStream fos = new FileOutputStream(SDCard_PATH);
            // 把 assets 下的图片拷入 sd 卡
            InputStream is = getResources().getAssets().open("image.jpg");
            byte buffer[] = new byte[8192]; // 每次最多写入 8k
            int count = 0;
            while ((count = is.read(buffer)) >= 0) {
                fos.write(buffer, 0, count);
```

```
            }
            fos.close();
            is.close();
Toast.makeText(context, "图片已经存入"+SDCard_PATH, Toast.LENGTH_SHORT).show();
        } catch (Exception e) {
            Toast.makeText(context, "写入失败！", Toast.LENGTH_SHORT).show();
        }
    }
```

（6）readFromSDCard()方法调用 getBitmap2()方法从指定路径读取图片，并给予对应操作的提示信息。

```
    private void readFromSDCard() {
        String name = SDCard_PATH;
        if (!new File(name).exists()) {
            Toast.makeText(context,android.os.Environment.getExternalStorageDirectory()+
                    "中没有图像", Toast.LENGTH_SHORT).show();
            return;
        }
        try {
            // 压缩载入图片
            Bitmap bitmap = getBitmap2(name, width, height);
            ivSD.setImageBitmap(bitmap);
        } catch (Exception e) {
            Toast.makeText(context, "载入图像失败！", Toast.LENGTH_SHORT).show();
        }
    }
```

（7）getBitmap2()方法通过计算原始图片的宽和高，然后压缩图片，最后得到压缩过的 bitmap 图片。

```
    public static Bitmap getBitmap2(String imageFilePath, int displayWidth,
            int displayHeight) {
        BitmapFactory.Options bitmapOptions = new BitmapFactory.Options();
        bitmapOptions.inJustDecodeBounds = true;
        Bitmap bmp = BitmapFactory.decodeFile(imageFilePath, bitmapOptions);
```

```
// 编码后 bitmap 的宽高,bitmap 除以屏幕宽度得到压缩比
int widthRatio = (int) FloatMath.ceil(bitmapOptions.outWidth
                                                    / (float) displayWidth);
int heightRatio = (int) FloatMath.ceil(bitmapOptions.outHeight
                                                    / (float) displayHeight);

if (widthRatio > 1 && heightRatio > 1) {
    if (widthRatio > heightRatio) {
        // 压缩到原来的(1/widthRatios)
        bitmapOptions.inSampleSize = widthRatio;
    } else {
        bitmapOptions.inSampleSize = heightRatio;
    }
}
bitmapOptions.inJustDecodeBounds = false;
bmp = BitmapFactory.decodeFile(imageFilePath, bitmapOptions);
return bmp;
    }
}
```

4. 扩展练习

（1）修改实验程序，在 Android 开发终端的内存中创建一个（自定义名字）.txt 文件，并写入内容，读取出文件的内容，并输出文件内容。在 LogCat 中查看输出消息，如图 16-4 和图 16-5 所示。

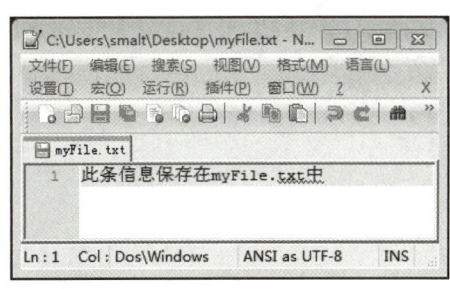

图 16-4　新建 txt 文件

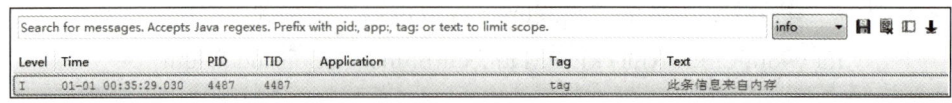

图 16-5　打印出内存中 txt 文件的内容

（2）修改实验程序，在 Android 开发终端的 SD 卡中创建一个（自定义名字）.txt 文件，并写入内容，读取出文件内容，并输出文件内容。在 LogCat 中查看输出消息，如图 16-6、图 16-7 所示。

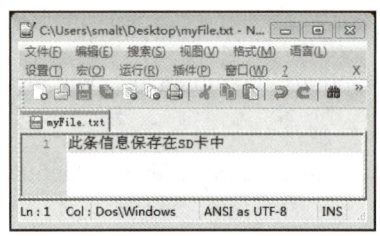

图 16-6　sd 卡中的 txt 文件

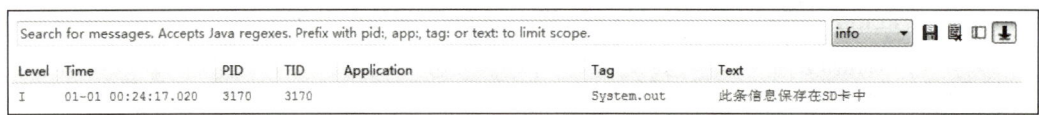

图 16-7　打印出 sd 卡中 txt 文件内容

5．项目验证

用 Android 开发终端连接电脑，在电脑中启动 eclipse 并运行实验项目程序。

（1）单击程序图标进入程序，在"读写内存中的文件"栏下单击【写入】按钮，然后在弹出的对话框中输入要写入内存的字符串，如图 16-8 所示，单击【确定】按钮，看是否成功写入数据，如图 16-9 所示，然后单击【读取】按钮读取写入的信息，如图 16-10 所示。

项目十六 IO 操作与数据存储访问

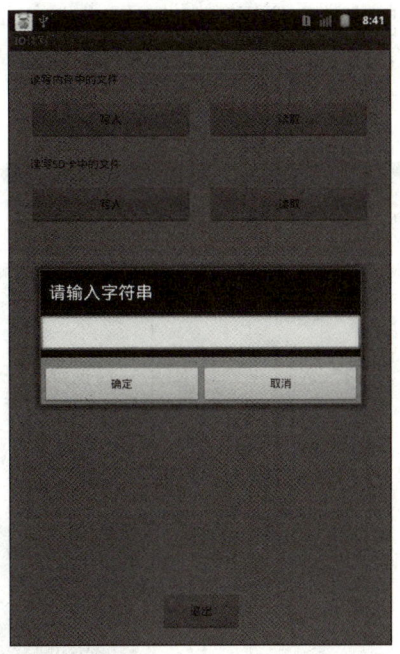

图 16-8　输入字符串

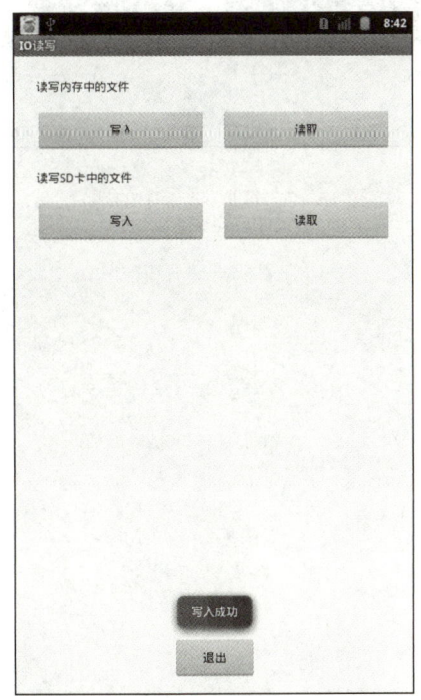

图 16-9　写入内存成功

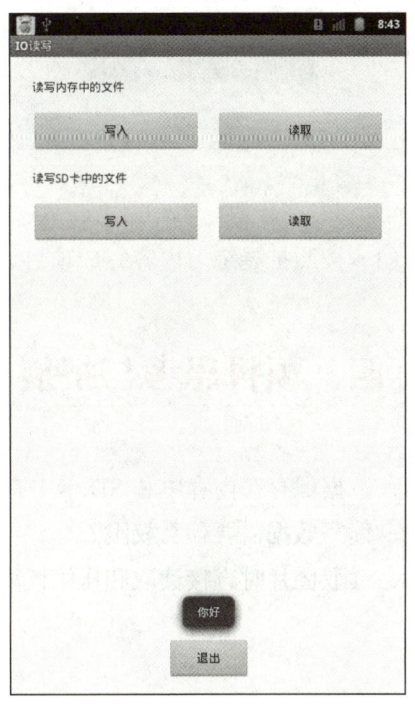

图 16-10　读取写入内存的信息

（2）在"读写 SD 卡中的文件"栏下单击"写入"按钮把指定图片存入 SD 卡中，查看是否写入成功，如图 16-11 所示，然后单击"读取"按钮读取写入的图片，如图 16-12 所示。

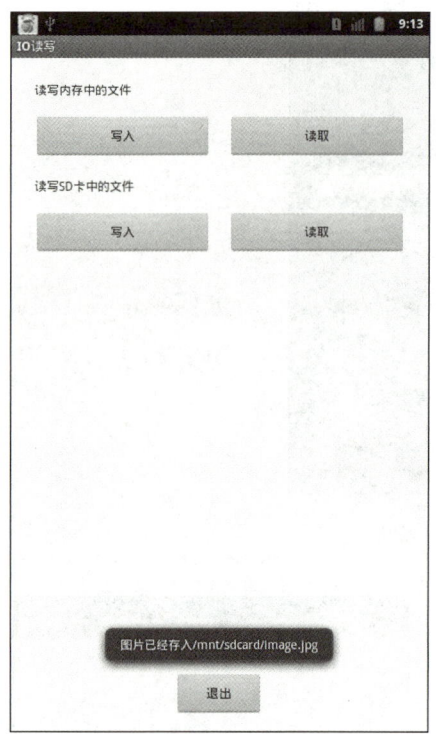
图 16-11　保存读入的图片

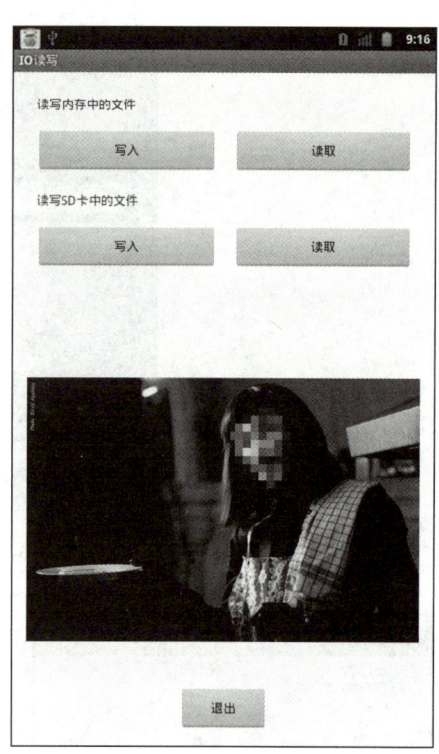
图 16-12　读取 sd 卡中的图片

（3）按照扩展练习内容完成项目调试。

四、项目思考与扩展

1．数据保存在内存中和 SD 卡中有何不同？
2．读写数据需要什么权限？
3．读取图片时直接读取和压缩读取有何区别？

项目十七　SQLite 实现数据的存储与访问

【本章导读】

　　Android 系统集成了一个轻量级的数据库——SQLite，SQLite 并不想成为像 Oracle、MySQL 那样的数据库。SQLite 只是一个嵌入式的数据库引擎，专门适用于资源有限的设备上（如手机、PDA 等）适量的数据存取。

　　虽然 SQLite 支持绝大部分 SQL 92 语法，也允许开发者使用 SQL 语句操作数据库中的数据，但 SQLite 并不像 Oracle、MySQL 数据库那样需要安装、启动服务器进程，SQLite 数据库只是一个文件。

一、项目要求

1. 掌握 SQLite 创建数据库的方法。
2. 掌握 SQLite 中数据的添加、修改、删除操作，熟悉 SQLite 的操作。
3. 使用 SQLite 实现图书信息管理程序，完成图书信息的录入、查询、修改和删除等操作。

二、项目相关知识

1. SQLite 简介

SQLite 是遵守 ACID 的关系数据库管理系统，它包含在一个相对小的 C 库，是 D.RichardHipp 建立的公有领域项目。

不像常见的客户端/服务器结构范例，SQLite 引擎不是一个程序与之通信的独立进程，而是连接到程序中成为它的一个主要部分。所以主要的通信协议是在编程语言内的直接 API 调用。这在消耗总量、延迟时间和整体简单性上有积极的作用。整个数据库（定义、表、索引和数据本身）都在宿主主机上存储在一个单一的文件中。它简单的设计是通过在开始一个事务的时候锁定整个数据文件而完成的。

Android 提供了 SQLiteDatabase 代表一个数据库（底层就是一个数据库文件），一旦应用程序获得了代表指定数据库的 SQLiteDatabase 对象，接下来就可通过 SQLiteDatabase 对象来管理、操作数据库了。

SQLiteDatabase 提供了以下静态方法来打开一个文件对应的数据库：

（1）static SQLiteDatabase openDatabase (String path, SQLiteDatabase. CursorFactory factory, int flags)：打开 path 文件所代表的 SQLite 数据库。

（2）static SQLiteDatabase openOrCreateDatabase (File file, SQLiteDatabase. CursorFactory factory)：打开或创建（如果不存在）file 文件所代表的 SQLite 数据库。

（3）static SQLiteDatabase openOrCreateDatabase (String path, SQLiteDatabase. CursorFactory factory)：打开或创建（如果不存在）path 文件所代表的 SQLite 数据库。

在程序中获取 SQLiteDatabase 对象之后，接下来就可调用 SQLiteDatabase 的如下方法来操作数据库了：

（1）execSQL(String sql, Object[] bindArgs)：执行带占位符的 SQL 语句。

（2）execSQL(String sql)：执行 SQL 语句。

（3）insert(String table, String nullColumnHack, ContentValues values)：向执行表中插入数据。

（4）update(String table, ContentValues values, String whereClause, String[] whereArgs)：更新指定表中的特定数据。

（5）delete(String table, String whereClause, String[] whereArgs)：删除指定表中的特定数据。

（6）Cursor query(String table, String[] columns, String whereClause, String[]whereArgs, String groupBy, String having, String orderBy)：对执行数据表执行查询。

（7）Cursor query(String table,String[] columns, String whereClause, String[]whereArgs, String groupBy, String having, String orderBy, String limit)：对执行数据表执行查询。limit 参数控制最多查询几条记录（用于控制分页的参数）。

（8）Cursor query(boolean distinct, String table, String[] columns, String whereClause, String[] whereArgs, String groupBy, String having, String orderBy, String limit)：对指定表执行查询语句。其中第一个参数控制是否去除重复值。

（9）rawQuery(String sql, String[] selectionArgs)：执行带占位符的 SQL 查询。

（10）beginTransaction()：开始事务。

（11）endTransaction()：结束事务。

从上面的方法不难看出，其实 SQLiteDatabase 的作用有点类似于 JDBC 的 Connection 接口，但 SQLiteDatabase 提供的方法更多：比如 insert、update、delete、query 等方法，其实这些方法完全可通过执行 SQL 语句来完成，但 Android 考虑到部分开发者对 SQL 语法不熟悉，所以提供这些方法帮助开发者以更简单的方式来操作数据表的数据。

上面的查询方法都是返回一个 Cursor 对象，Android 中的 Cursor 类似于 JDBC 的 ResultSet，Cursor 同样提供了如下方法来移动查询结果的记录指针。

（1）move(int offset)：将记录指针向上或向下移动指定的行数。offset 为正数就是向下移动；为负数就是向上移动。

（2）boolean moveToFirst()：将记录指针移动到第一行，如果移动成功则返回 true。

（3）boolean moveToLast()：将记录指针移动到最后一行，如果移动成功则返回 true。

（4）boolean moveToNext()：将记录指针移动到下一行，如果移动成功则返回 true。

（5）boolean moveToPosition(int position)：将记录指针移动到指定的行，如果移动成功则返回 true。

（6）bonlean moveToPrevious()：将记录指针移动到上一行，如果移动成功则返回 true。

一旦将记录指针移动到指定行之后，接下来就可以调用 Cursor 的 getXxx()方法获取该行的指定列的数据。

2. SQLite 对数据库的操作

操作数据一定会用到 SQLiteOpenHelper 类，写一个工具类继承 SQLiteOpenHelper 类，在构造函数里创建数据库：

```
public BookDao(Context context) {
    uper(context, DB_NAME, null，DB_VERSION);
}
```

在 onCreate()方法里创建表：

```
public void onCreate(SQLiteDatabase db) {
    og.i("tag", "onCreate");
    tring sql = "create table " + TAB_NAME + " (" + BookBean.FILED_ID+
        " integer primary key autoincrement, "+ BookBean.FILED_BOOKNAME +
        " varchar(20), "+ BookBean.FILED_WRITER+ " text, " +
    BookBean.FILED_PRESS + " varchar(20), "+ BookBean.FILED_PRICE + " long)";
    System.out.println("sql--->" + sql);
    db.execSQL(sql);
}
```

(1) 在表中增加数据用到如下代码：

```
String sql = "insert into " + BookDao.TAB_NAME+
                            "(bookname, writer, press, price) values(?, ?, ?, ?)";
db.execSQL(sql, new Object[] { "演员的自我修养", "周星驰", "银河映像", 23.8 });
Log.i("tag", "插入成功!");
```

(2) 在表中删除数据用到如下代码：

```
db.delete(BookDao.TAB_NAME, "writer=?", new String[] { "周星驰" });
```

(3) 在表中修改数据用到如下代码：

```
SQLiteDatabase db = helper.getWritableDatabase();
ContentValues values = new ContentValues();
values.put("name", name.getText().toString());
values.put("age", new Integer(age.getText().toString()) );
values.put("email", email.getText().toString());
values.put("address", address.getText().toString());
db.update(DatabaseHelper.TABLE_NAME, values,
//要修改的值存在这里  "_id=?", //查询条件  new String[]{personId});
```

(4) 在表中查询数据用到如下代码：

```
Cursor c = db.rawQuery("select * from " + BookDao.TAB_NAME, null);
    while (c.moveToNext()) {
        System.out.println("-->" + c.getInt(0) + ", "
            + c.getString(1) + ", " + c.getString(2) + ", "
            + c.getString(3) + ", " + c.getString(4));
}
```

(5) 模糊查询为：

String sql = "select * from lib_book_tab where writer like '%星%'";

3. 动态广播的使用

定义广播接收器：

```java
class SendReceicer extends BroadcastReceiver {
    @Override
    public void onReceive(Context context, Intent intent) {
        // TODO Auto-generated method stub
        if (intent.getAction().equals("update")) {
            System.out.println("have receive...");
        }
    }
}
```

在 onCreate() 中注册：

```java
intentFilter = new IntentFilter();
intentFilter.addAction("update");
receiver = new SendReceicer();
this.registerReceiver(receiver, intentFilter);
```

在 onDestroy() 中注销：

```java
@Override
protected void onDestroy() {
    this.unregisterReceiver(receiver);
    super.onDestroy();
}
```

发送：

```java
Intent intent = new Intent();
intent.setAction("update");
SendActivity.this.sendBroadcast(intent);
System.out.println("send receiver");
}
```

三、项目实施过程

1. 工程创建

参考前面实验项目中创建工程的步骤，在 eclipse 中新建一个名为 AndroidCode17 的 Android 项目，工程目录结构如图 17-1 所示。

2. XML 布局文件的开发

主界面布局如图 17-2 所示，总体布局采用 RelativeLayout，上部是一个 TextView 显示标题，中部一个 TableLayout 布局包括 4 个 TextView 和 4 个 EditText 显示图书信息，下部是 3 个 Button 按钮，分别是【查询】【增加】和【退出】。

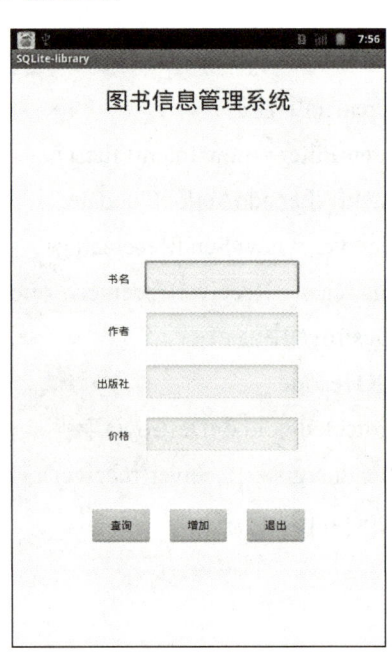

图 17-1　工程目录结构图　　　　　　　图 17-2　主界面布局

（1）主界面输入部分由 4 个 EditText 组成，输入标题由 4 个 TextView 组成，代码如下：

```
<?xml version="1.0" encoding="utf-8"?>
<RelativeLayout xmlns:android="http://schemas.android.com/apk/res/android"
    android:layout_width="match_parent"
    android:layout_height="match_parent" >
```

```xml
<!-- 标题部分 -->
<TextView
        android:layout_width="wrap_content"
        android:layout_height="wrap_content"
        android:layout_alignParentTop="true"
        android:layout_centerHorizontal="true"
        android:layout_marginTop="20dp"
        android:text="@string/str_titile"
        android:textSize="30sp" />
<!-- 编辑部分 -->
<TableLayout
        android:id="@+id/main_layout_mid"
        android:layout_width="wrap_content"
        android:layout_height="wrap_content"
        android:layout_centerHorizontal="true"
        android:layout_centerVertical="true" >
<!-- 书名 -->
<TableRow>
        <TextView
                android:layout_margin="10dp"
                android:gravity="center|right"
                android:text="@string/str_bookname" />
        <EditText
                android:id="@+id/main_bookname"
                android:layout_width="200dp"
                android:layout_margin="10dp"
                android:inputType="text" />
</TableRow>
<!-- 作者 -->
<TableRow>
        <TextView
                android:layout_margin="10dp"
                android:gravity="center|right"
```

```xml
            android:text="@string/str_writer" />
        <EditText
            android:id="@+id/main_writer"
            android:layout_width="200dp"
            android:layout_margin="10dp"
            android:inputType="text" />
</TableRow>
<!-- 出版社 -->
<TableRow>
    <TextView
        android:layout_margin="10dp"
        android:gravity="center|right"
        android:text="@string/str_press" />
    <EditText
        android:id="@+id/main_press"
        android:layout_width="200dp"
        android:layout_margin="10dp"
        android:inputType="text" />
</TableRow>
<!-- 价格 -->
<TableRow>
    <TextView
        android:layout_margin="10dp"
        android:gravity="center|right"
        android:text="@string/str_price" />
    <EditText
        android:id="@+id/main_price"
        android:layout_width="200dp"
        android:layout_margin="10dp"
        android:inputType="numberDecimal" />
</TableRow>
</TableLayout>
<!-- 按钮部分 -->
```

```xml
<TableLayout
    android:layout_width="wrap_content"
    android:layout_height="wrap_content"
    android:layout_below="@id/main_layout_mid"
    android:layout_centerHorizontal="true"
    android:layout_marginTop="20dp" >
<TableRow>
    <Button
        android:id="@+id/main_btn_query"
        android:layout_width="80dp"
        android:layout_margin="10dp"
        android:gravity="center"
        android:text="@string/btn_query" />
    <Button
        android:id="@+id/main_btn_add"
        android:layout_width="80dp"
        android:layout_margin="10dp"
        android:gravity="center"
        android:text="@string/btn_add" />
    <Button
        android:id="@+id/main_btn_exit"
        android:layout_width="80dp"
        android:layout_margin="10dp"
        android:gravity="center"
        android:text="@string/str_exit" />
</TableRow>
</TableLayout>
</RelativeLayout>
```

（2）查询结果界面由布局界面（result_layout.xml）和选项界面（book_item）组成，result_layout.xml 界面包括一个 ListView 显示图书项目和一个 Button 返回按钮，book_item 界面包括 5 个 TextView 和 5 个 EditText，分别显示图书的编号、书名、作者、出版社和售价，整体布局如图 17-3 所示。

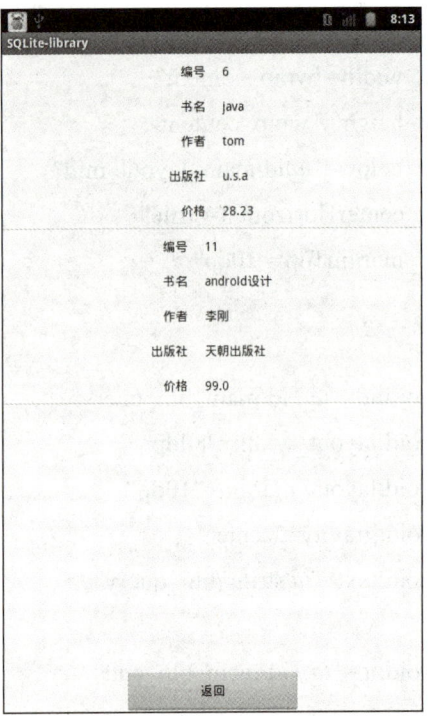

图 17-3　查询结果显示界面

代码如下：

```xml
<?xml version="1.0" encoding="utf-8"?>
<RelativeLayout xmlns:android="http://schemas.android.com/apk/res/android"
    android:layout_width="match_parent"
    android:layout_height="match_parent" >
<!-- 查询结果界面 -->
<Button
        android:id="@+id/res_ly_btn_back"
        android:layout_width="200dp"
        android:layout_height="wrap_content"
        android:layout_alignParentBottom="true"
        android:layout_centerHorizontal="true"
        android:layout_margin="10dp"
        android:text="@string/str_back" />
<ListView
        android:id="@+id/listview"
```

```
            android:layout_width="fill_parent"
            android:layout_height="fill_parent"
            android:layout_above="@id/res_ly_btn_back" >
    </ListView>
</RelativeLayout>
```

（3）修改界面如图 17-4 所示，上部是一个 TableLayout 布局，包括 4 个 TextView 和 4 个 EditText 按钮，分别表示图书名、作者、出版社和价格，下方是 TableLayout 布局的两个 Button，代表【确定】和【取消】按钮。

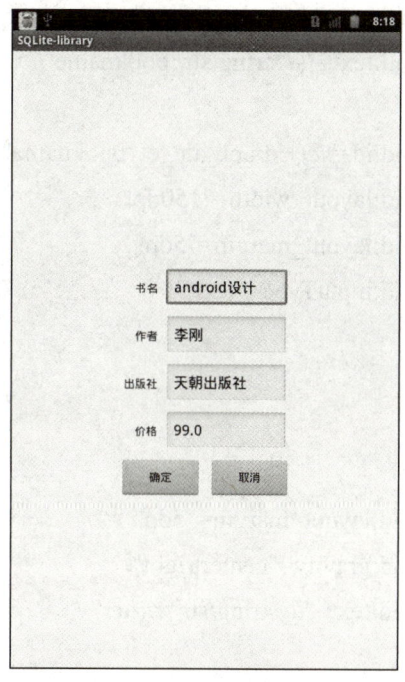

图 17-4　修改图书信息界面

代码如下：

```
<?xml version="1.0" encoding="utf-8"?>
<RelativeLayout xmlns:android="http://schemas.android.com/apk/res/android"
    android:id="@+id/update_item_ly"
    android:layout_width="match_parent"
    android:layout_height="match_parent"
    android:background="#ffffffff" >
    <TableLayout
        android:id="@+id/upd_tbly_top"
```

```xml
            android:layout_width="wrap_content"
            android:layout_height="wrap_content"
            android:layout_centerHorizontal="true"
            android:layout_centerVertical="true" >
<!-- 书名 -->
<TableRow>
    <TextView
                android:layout_margin="5dp"
                android:gravity="center|right"
                android:text="@string/str_bookname" />
        <EditText
                android:id="@+id/upd_ite_et_bookname"
                android:layout_width="150dp"
                android:layout_margin="5dp"
                android:inputType="text" />
</TableRow>
<!-- 作者 -->
<TableRow>
        <TextView
                android:layout_margin="5dp"
                android:gravity="center|right"
                android:text="@string/str_writer" />
        <EditText
                android:id="@+id/upd_iem_et_wrtier"
                android:layout_width="150dp"
                android:layout_margin="5dp"
                android:inputType="text" />
</TableRow>
<!-- 出版社 -->
<TableRow>
        <TextView
                android:layout_margin="5dp"
                android:gravity="center|right"
```

```xml
            android:text="@string/str_press" />
        <EditText
            android:id="@+id/upd_iem_et_press"
            android:layout_width="150dp"
            android:layout_margin="5dp"
            android:inputType="text" />
    </TableRow>
    <!-- 价格 -->
    <TableRow>
        <TextView
            android:layout_margin="5dp"
            android:gravity="center|right"
            android:text="@string/str_price" />
        <EditText
            android:id="@+id/upd_iem_et_price"
            android:layout_width="150dp"
            android:layout_margin="5dp"
            android:inputType="text" />
    </TableRow>
</TableLayout>
<!-- 确定和取消按钮 -->
<TableLayout
    android:layout_width="wrap_content"
    android:layout_height="wrap_content"
    android:layout_below="@id/upd_tbly_top"
    android:layout_centerHorizontal="true" >
    <TableRow>
        <Button
            android:id="@+id/upd_btn_ok"
            android:layout_width="100dp"
            android:layout_margin="5dp"
            android:text="@string/str_ok" />
        <Button
```

```
                    android:id="@+id/upd_btn_cancel"
                    android:layout_width="100dp"
                    android:layout_margin="5dp"
                    android:text="@string/str_cancel" />
        </TableRow>
    </TableLayout>
</RelativeLayout>
```

3. Java 文件的开发

程序分为 6 个 class 文件，BookBean 是图书的 JavaBean 类（JavaBean 是一种 JAVA 语言写成的可重用组件），BookDao 是数据库操作类，BookAdapter 是 ListView 的显示适配器类，QueryActivity 是主界面，ResultActivity 是查询结果显示类，UpdateActivity 是修改图书类。程序代码结构比较清晰，下面对关键代码进行说明。

（1）BookBean 中包括图书的编号、书名、作者、出版社和价格五个属性，并设置公用静态常量和 get、set 方法为接口供其他类使用。

```java
public class BookBean {
    // 常量***********************************
    public final static String FILED_ID = "_id";
    public final static String FILED_BOOKNAME = "bookname";
    public final static String FILED_WRITER = "writer";
    public final static String FILED_PRESS = "press";
    public final static String FILED_PRICE = "price";
    /**
     * id 号
     */
    private int id;
    /**
     * 书名
     */
    private String bookname;
    /**
     * 作者
     */
```

```java
private String writer;
/**
 * 出版社
 */
private String press;
/**
 * 价格
 */
private double price;

/**
 * @description 构造函数
 * @return
 */
public BookBean(int id, String bookname, String writer, String press,
        double price) {
    this.id = id;
    this.bookname = bookname;
    this.writer = writer;
    this.press = press;
    this.price = price;
}

public int getId() {
    return id;
}

public void setId(int id) {
    this.id = id;
}

public String getBookname() {
    return bookname;
```

```java
    }

    public void setBookname(String bookname) {
        this.bookname = bookname;
    }

    public String getWriter() {
        return writer;
    }

    public void setWriter(String writer) {
        this.writer = writer;
    }

    public String getPress() {
        return press;
    }

    public void setPress(String press) {
        this.press = press;
    }

    public double getPrice() {
        return price;
    }

    public void setPrice(double price) {
        this.price = price;
    }
}
```

（2）BookDao 类提供了建立数据库的方法，并在 onCreate 方法中调用 SQL 命令建立数据库。

```java
public class BookDao extends SQLiteOpenHelper {
    /**
     * 数据库名字
     */
    public static String DB_NAME = "lib.db";
    /**
     * 表名
     */
    public static String TAB_NAME = "lib_book_tab";
    /**
     * 版本号
     */
    public static int DB_VERSION = 1;
    /**
     * bookbean
     */
    public BookDao(Context context) {
        super(context, DB_NAME, null, DB_VERSION);
    }

    /**
     * @description 创建数据库
     * @param context
     *     内容
     * @param name
     * 数据库名字
     * @param factory
     * CursorFactory 指定在执行查询时获得一个游标实例的工厂类,
     *设置为 null 代表使用系统默认的工厂类
     * @param version
     *        版本号
     */
    public BookDao(Context context, String name, CursorFactory factory,int version) {
```

```java
        super(context, DB_NAME, factory, DB_VERSION);
        // TODO Auto-generated constructor stub
    }

    /**
     * @description 建表
     */
    @Override
    public void onCreate(SQLiteDatabase db) {
        Log.i("tag", "onCreate");
        String sql = "create table " + TAB_NAME + " (" + BookBean.FILED_ID
                + " integer primary key autoincrement, "
                + BookBean.FILED_BOOKNAME + " varchar(20), "
                + BookBean.FILED_WRITER
                + " text, " + BookBean.FILED_PRESS + " varchar(20), "
                + BookBean.FILED_PRICE + " long)";
        System.out.println("sql--->" + sql);
        db.execSQL(sql);
    }
    /**
     * @description 更新数据库
     */
    @Override
    public void onUpgrade(SQLiteDatabase db, int oldVersion, int newVersion) {
        Log.i("tag", "onUpgrade");
        String sql = "drop table if exist " + TAB_NAME;
        db.execSQL(sql);
        onCreate(db);
    }
    /**
     * @description 插入
     */
    public void insert(){
```

```
//      SQLiteDatabase db=DB_NAME;
    }
}
```

（3）BookAdapter 类设置了自定义的适配器，使得在 ListView 中能够显示指定的数据，下面代码中 getView()为获取每条图书信息的关键方法。

```
public class BookAdapter extends BaseAdapter {
    @SuppressWarnings("unused")
    private Context context;
    private LayoutInflater inflater; // 视图容器
    private List<Map<String, Object>> listItem;
    private String index[];

    public final class ViewHolder {
        public TextView tvId, tvBookname, tvWriter, tvPress, tvPrice;
    }

    /**
     * @description 创建视图设置上下文
     */
    public BookAdapter(Context context, List<Map<String, Object>> list,
            String index[]) {
        this.index = index;
        this.context = context;
        inflater = LayoutInflater.from(context);
        this.listItem = list;
    }

    /**
     * @description 子项目个数
     */
    @Override
    public int getCount() {
        // TODO Auto-generated method stub
```

```java
            return listItem.size();
        }

        @Override
        public Object getItem(int position) {
            // TODO Auto-generated method stub
            return listItem.get(position);
        }

        @Override
        public long getItemId(int position) {
            // TODO Auto-generated method stub
            return position;
        }

        @Override
        public View getView(int position, View convertView, ViewGroup parent) {
            // TODO Auto-generated method stub
            ViewHolder holder = null;
            if (convertView == null) {
                holder = new ViewHolder();
                String strId = listItem.get(position).get(index[0]) + "";
                String strBookname = listItem.get(position).get(index[1]) + "";
                String strWriter = listItem.get(position).get(index[2]) + "";
                String strPress = listItem.get(position).get(index[3]) + "";
                String strPrice = listItem.get(position).get(index[4]) + "";

                // 获取文件视图
                convertView = inflater.inflate(R.layout.book_item, null);
                holder.tvId = (TextView) convertView
                        .findViewById(R.id.book_item_id);
                holder.tvBookname = (TextView) convertView
                        .findViewById(R.id.book_item_bookname);
```

```
            holder.tvWriter = (TextView) convertView
                    .findViewById(R.id.book_item_writer);
            holder.tvPress = (TextView) convertView
                    .findViewById(R.id.book_item_press);
            holder.tvPrice = (TextView) convertView
                    .findViewById(R.id.book_item_price);

            // 设置字符串内容
            holder.tvId.setText(strId);
            holder.tvBookname.setText(strBookname);
            holder.tvWriter.setText(strWriter);
            holder.tvPress.setText(strPress);
            holder.tvPrice.setText(strPrice);

            // 控件到 converview
            convertView.setTag(holder);
        } else {
            holder = (ViewHolder) convertView.getTag();
        }
        return convertView;
    }
}
```

（4）QueryActivity 类是查询图书的主程序，其中 onCreate()方法在程序启动时候调用，initListener()方法给控件注册监听器，initView()方法初始化控件并和 XML 文件中绘制的控件建立联系，使得单击不同按钮调用相应的单击事件。queryData()方法用于查询输入的图书信息，程序获取 EditText 中输入的内容通过 Intent 信使传递给 ResultActivity。

```
public class QueryActivity extends Activity implements OnClickListener {
    // 控件
    EditText etBookname, etWriter, etPress, etPrice;
    Button btnQuery, btnAdd, btnExit;
    // 常量
    public final static String BUNDLE = "bundle";
```

```java
@Override
protected void onCreate(Bundle savedInstanceState) {
    // TODO Auto-generated method stub
    super.onCreate(savedInstanceState);
    setContentView(R.layout.main);
    // 初始化控件
    initView();
    // 设置监听
    initListener();
}
private void initListener() {
    btnAdd.setOnClickListener(this);
    btnExit.setOnClickListener(this);
    btnQuery.setOnClickListener(this);
}
private void initView() {
    etBookname = (EditText) findViewById(R.id.main_bookname);
    etWriter = (EditText) findViewById(R.id.main_writer);
    etPress = (EditText) findViewById(R.id.main_press);
    etPrice = (EditText) findViewById(R.id.main_price);
    btnAdd = (Button) findViewById(R.id.main_btn_add);
    btnQuery = (Button) findViewById(R.id.main_btn_query);
    btnExit = (Button) findViewById(R.id.main_btn_exit);
}

@Override
public void onClick(View v) {
    String strBookname = etBookname.getText().toString().replace(" ", ""), strWriter = etWriter
            .getText().toString().replace(" ", ""), strPress = etPress
            .getText().toString().replace(" ", ""), strPrice = etPrice
            .getText().toString().replace(" ", "");
    switch (v.getId()) {
    case R.id.main_btn_query:
```

```
            // 查询弹出结果界面显示,传递参数
            queryData(strBookname, strWriter, strPress, strPrice);
            break;
        case R.id.main_btn_exit:
            finish();
            break;
        case R.id.main_btn_add:
            // 添加新记录
            addData(strBookname, strWriter, strPress, strPrice);
            break;
        default:
            break;
        }
    }
}
```

addData()方法使得在 EditText 中输入的数据存入数据库，调用数据库查询命令 db.execSQL()方法执行操作。

```
private void addData(String strBookname, String strWriter, String strPress,
        String strPrice) {
    if (strBookname.equals("") || strWriter.equals("")
            || strPress.equals("") || strPrice.equals("")) {
        Toast.makeText(this, "请输入完整信息!", Toast.LENGTH_SHORT).show();
    } else {
        BookDao dao = new BookDao(this);
        SQLiteDatabase db = dao.getWritableDatabase();
        try {
            String sql = "insert into " + BookDao.TAB_NAME
                    + "(bookname,writer,press,price) values(?,?,?,?)";
            db.execSQL(sql, new Object[] { strBookname, strWriter,
                    strPress, Double.parseDouble(strPrice) });
            Toast.makeText(this, "添加成功!", Toast.LENGTH_SHORT).show();
        } catch (Exception e) {
            Toast.makeText(this, "添加失败!请重试!", Toast.LENGTH_SHORT).show();
```

```
                    e.printStackTrace();
                }
            }
        }
```

（5）ResultActivity 类显示查询结果，onCreate()方法在程序创建时通过 initView()和 listener()方法初始化控件建立和 XML 文件绘制控件的联系以及注册监听事件来响应不同控件的单击事件。

getItemList()方法用于获取图书信息列表的 list 对象：

```
        private ArrayList<BookBean> getItemList() {
            ArrayList<BookBean> list = new ArrayList<BookBean>();
            dao = new BookDao(ResultActivity.this);
            db = dao.getWritableDatabase();
            // 查询结果存入 list
            try {
                // 获取传递过来的参数作为查询限制
                Intent intent = getIntent();
                Bundle bundle = intent.getBundleExtra(QueryActivity.BUNDLE);
                String strBookname = bundle.getString(BookBean.FILED_BOOKNAME);
                String strWriter = bundle.getString(BookBean.FILED_WRITER);
                String strPress = bundle.getString(BookBean.FILED_PRESS);
                String strPrice = bundle.getString(BookBean.FILED_PRICE);
                String sql = "select * from " + BookDao.TAB_NAME + " where ";
                ArrayList<String> list1 = new ArrayList<String>();
                if (!strBookname.equals("")) {
list1.add(BookBean.FILED_BOOKNAME + " like '%" + strBookname+ "%'");
                }
                if (!strWriter.equals("")) {
                    list1.add(BookBean.FILED_WRITER + " like '%" + strWriter + "%'");
                }
                if (!strPress.equals("")) {
                    list1.add(BookBean.FILED_PRESS + " like '%" + strPress + "%'");
                }
```

```java
            if (!strPrice.equals("")) {
                list1.add(BookBean.FILED_PRICE + " = " + strPrice);
            }
            // sql 查询语句拼接
            String str1 = "";
            if (list1.size() > 1) {
                for (int i = 0; i < list1.size() - 2; i++) {
                    str1 += " " + list1.get(0) + " and ";
                }
                str1 += " " + list1.get(list1.size() - 1);
            } else {
                str1 = list1.get(0);
            }
            sql += str1;
            System.out.println("sql-->" + sql);
            Cursor c = db.rawQuery(sql, null);
            while (c.moveToNext()) {
                // 存入数据集
                BookBean book = new BookBean(c.getInt(0), c.getString(1),
                        c.getString(2), c.getString(3), c.getDouble(4));
                list.add(book);
            }
            Log.i("tag", "查找成功!");
        } catch (Exception e) {
            Log.i("tag", "查找失败!");
            e.printStackTrace();
        }
        return list;
    }
```

getDataMap 方法用于获取图书列表的 map 映射，将图书数据一一对应到 ListView 列表中显示出来。

```java
    private List<Map<String, Object>> getDataMap(ArrayList<BookBean> resultList) {
```

```java
        List<Map<String, Object>> dataMap = new ArrayList<Map<String, Object>>();
        Map<String, Object> map;
        for (BookBean book : resultList) {
            map = new HashMap<String, Object>();
            map.put("id", book.getId() + "");
            map.put("bookname", book.getBookname());
            map.put("writer", book.getWriter());
            map.put("press", book.getPress());
            map.put("price", book.getPrice() + "");
            dataMap.add(map);
        }
        return dataMap;
    }
```

deleteData()方法用于删除选中的图书项，调用 db.delete()方法从数据库删除指定的图书记录信息。

```java
    private void deleteData() {
        try {
            db.delete(BookDao.TAB_NAME, BookBean.FILED_ID + "=?",
                    new String[] { idData });
            Toast.makeText(ResultActivity.this, "删除成功!", Toast.LENGTH_SHORT)
                    .show();
        } catch (Exception e) {
            Toast.makeText(ResultActivity.this, "删除失败!", Toast.LENGTH_SHORT)
                    .show();
            e.printStackTrace();
        }
        // 更新 listview
        listAdapter.remove(positionData);
        adapter.notifyDataSetChanged();
    }
```

（6）UpdateActivity 类用于修改选中的图书记录。其中 initView()方法初始块控件并通过 findViewById()方法建立与 XML 文件的联系，initListener()方法设置监听，当单击不同控件时调用对应的执行代码。在接收 Intent 传递的数据后，调用 update()方法修改数据库中的记录。

```java
    private void initListener() {
        btnOK.setOnClickListener(this);
        btnCancel.setOnClickListener(this);
    }
    private void initView() {
        etBookname = (EditText) findViewById(R.id.upd_ite_et_bookname);
        etWriter = (EditText) findViewById(R.id.upd_iem_et_wrtier);
        etPress = (EditText) findViewById(R.id.upd_iem_et_press);
        etPrice = (EditText) findViewById(R.id.upd_iem_et_price);
        btnOK = (Button) findViewById(R.id.upd_btn_ok);
        btnCancel = (Button) findViewById(R.id.upd_btn_cancel);
        // 获取 id 号,书名.作者,出版社,价格
        Intent intent = getIntent();
        Bundle bundle = intent.getBundleExtra("bundle");
        strID = bundle.getString(BookBean.FILED_ID);
        // 默认显示原来的数据

etBookname.setText(bundle.getString(BookBean.FILED_BOOKNAME).toString());
        etWriter.setText(bundle.getString(BookBean.FILED_WRITER).toString());
        etPress.setText(bundle.getString(BookBean.FILED_PRESS).toString());
        etPrice.setText(bundle.getString(BookBean.FILED_PRICE).toString());
    }
    public void onClick(View v) {
        switch (v.getId()) {
        case R.id.upd_btn_ok:
            String strBookname = etBookname.getText().toString();
            String strWriter = etWriter.getText().toString();
            String strPress = etPress.getText().toString();
            String strPrice = etPrice.getText().toString();
            // 修改记录
            try {
                BookDao dao = new BookDao(UpdateActivity.this);
```

```
                    SQLiteDatabase db = dao.getWritableDatabase();
                    ContentValues values = new ContentValues();
values.put(BookBean.FILED_BOOKNAME, strBookname);// key 为字段名，value 为值
values.put(BookBean.FILED_WRITER, strWriter);           // key 为字段名，value 为值
values.put(BookBean.FILED_PRESS, strPress);            // key 为字段名，value 为值
values.put(BookBean.FILED_PRICE, strPrice);            // key 为字段名，value 为值
db.update(BookDao.TAB_NAME, values, "_id=?",new String[] { strID });
Toast.makeText(UpdateActivity.this, "修改成功!", Toast.LENGTH_SHORT).show();
                    // 发送广播通知刷新 adapter
                    System.out.println("发送广播...");
                    finish();
                } catch (Exception e) {
Toast.makeText(UpdateActivity.this, "修改失败!", Toast.LENGTH_SHORT).show();
                    e.printStackTrace();
                }
                break;
            case R.id.upd_btn_cancel:
                finish();
            default:
                break;
            }
        }
```

4. 扩展练习

（1）修改实验程序，修改数据库表结构的创建程序代码，在数据库表中增加一列（图书编号），如图 17-5 所示。

（2）参考实验程序的基础代码部分，在程序代码中加入图书编号的输入框，输入图书编号，并在查询书籍时显示，如图 17-6 所示。

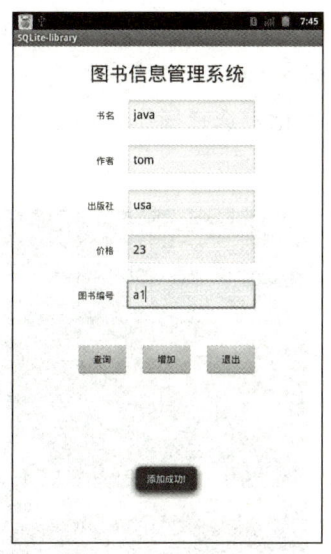

图 17-5 增加图书编号

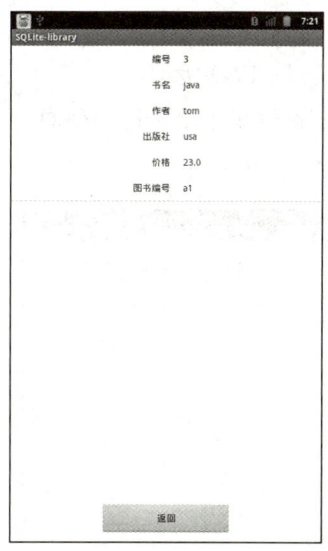

图 17-6 查询图书信息

5. 项目验证

（1）用 Android 开发终端连接电脑，在电脑中启动 Eclipse 并运行实验项目程序。程序运行后在主界面输入图书信息关键字，然后单击【查询】按钮可以查询到相关图书信息记录，其中"书名""作者""出版社"支持模糊查询，即只输入关键字即可查询到对应图书信息。例如查询书名包含"Java"的图书，在"书名"栏输入"Java"，如图 17-7 所示，然后单击【查询】按钮，程序将会跳转到结果显示界面，如图 17-8 所示。

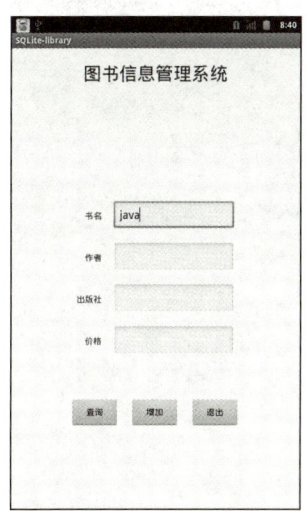

图 17-7 查询包含关键字"Java"的图书

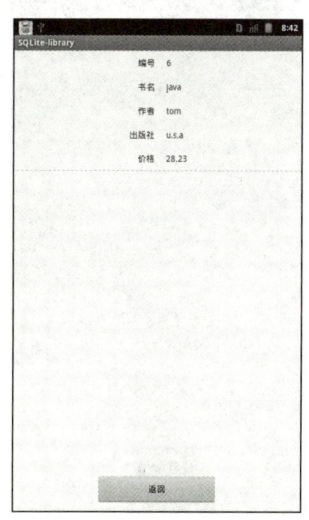

图 17-8 查询结果显示

（2）在主界面的文本框中输入图书信息数据，单击【增加】按钮则可以添加图书信息入数据库，如图 17-9 所示。

（3）在查询结果中单击图书信息项，会弹出对话框选择操作方式【删除】【修改】或【取消】，如图 17-10 所示。

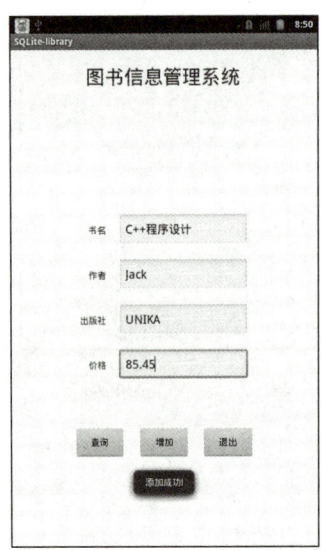 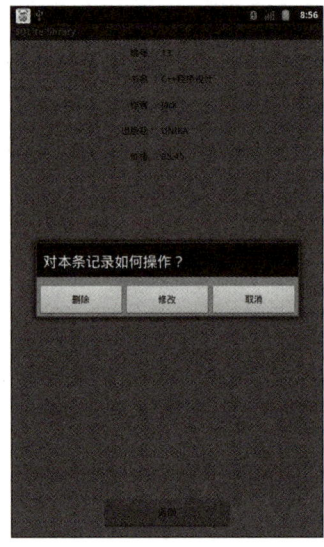

图 17-9　添加图书信息数据　　　　　图 17-10　对图书记录的操作

（4）单击【删除】选项可以删除选定的图书信息记录，如图 17-11 所示。

（5）若在下图界面选择【修改】可以进入修改图书信息界面，如图 17-12 所示。

 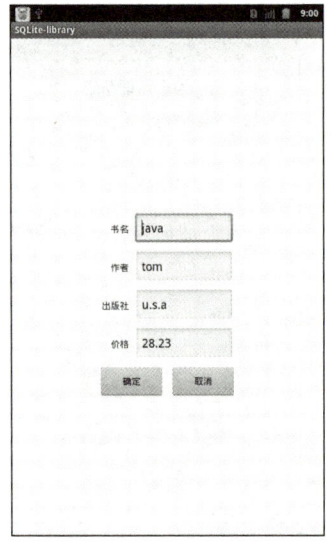

图 17-11　删除图书信息　　　　　图 17-12　修改图书信息界面

（6）填入需要修改的信息后，单击【确定】按钮可以完成对本条信息的修改，并在 ListView 中更新显示的信息，如图 17-13 所示。

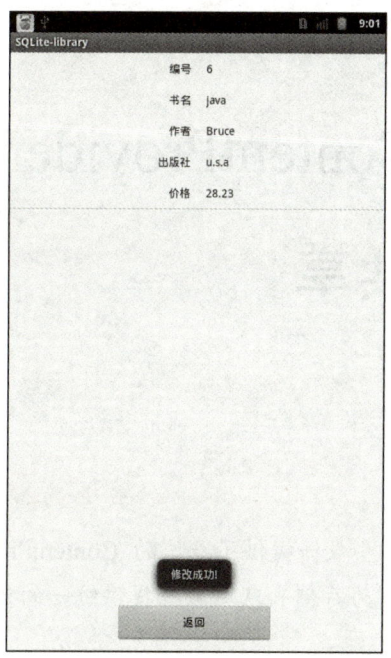

图 17-13　修改图书信息后的显示

（7）对照扩展练习给出的结果图示完成代码编程。

四、项目思考与扩展

1．SQLite 和其他数据库有何异同？
2．怎样高效的操作数据库和文件？
3．Intent 和 BroadcastReceiver 怎样组合才能高效？

项目十八　ContentProvider 实现数据共享

【本章导读】

Android 为常见的一些数据提供了默认的 ContentProvider，可以在不同的应用程序之间共享数据。它为存储和获取数据提供统一的接口。本节将以通讯录为例学习。

一、项目要求

1. 了解 ContentProvider 的实现原理。
2. 掌握使用 ContentProvider 对共享数据的查询、添加、删除操作。
3. 熟悉 ContentProvider 的 6 个重载函数。

二、项目相关知识

1. ContentProvider 介绍

在 Android 官方指出的 Android 的数据存储方式总共有五种，分别是：Shared Preferences、网络存储、文件存储、外储存储、SQLite。但是我们知道一般这些存储都只是在单独的一个应用程序之中达到一个数据的共享，有时候我们需要操作其他应用程序的一些数据，例如我们需要操作系统里的媒体库、通讯录等，这时我们就需要通过

ContentProvider 来满足需求了。

ContentProvider 是用来实现应用程序之间数据共享的类。当需要进行数据共享时，一般利用 ContentProvider 为需要共享的数据定义一个 URI，然后其他应用程序通过 Context 获得 ContentResolver 并将数据的 URI 传入即可。

Android 系统已经为一些常用的数据创建了 ContentProvider，这些 ContentProvider 都位于 android.provider 下，只要有相应的权限，自己开发的应用程序便可轻松地访问这些数据。

使用 ContentProvider 访问共享资源时，需要为应用程序添加适当的权限才可以。权限为<users-permission android:name="android.permission.READ_CONTACTS"/>。

对于 ContentProvider，最重要的就是数据模型（data model）和 URI。

1）数据模型

ContentProvider 为所有需要共享的数据创建一个数据表，在表中，每一行表示一条记录，而每一列代表某个数据，并且其中每一条数据记录都包含一个名为"_ID"的字段类标识每条数据。

2）URI

每个 ContentProvider 都会对外提供一个公开的 URI 来表示自己的数据集，当管理多个数据集时，将会为每个数据集分配一个独立的 URI。为系统的每一个资源确定一个名字，比方说通话记录。

（1）每一个 ContentProvider 都拥有一个公共的 URI，这个 URI 用于表示这个 ContentProvider 所提供的数据。

（2）Android 所提供的 ContentProvider 都存放在 android.provider 包中。将其分为 A，B，C，D 4 个部分，如图 18-1 所示。

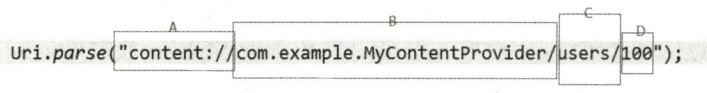

图 18-1 URI 路径

- A：标准前缀，用来说明一个 Content Provider 控制这些数据，无法改变的；"content://"。
- B：URI 的标识，用于唯一标识这个 ContentProvider，外部调用者可以根据这个标识来找到它。它定义了是哪个 Content Provider 提供这些数据。对于第三方应用程序，为了保证 URI 标识的唯一性，它必须是一个完整的、小写的类名。这个标识在元素的 authorities 属性中说明：一般是定义该 ContentProvider 的包类的名称。
- C：路径（path），通俗地讲就是你要操作的数据库中表的名字，或者你也可以自己定义，记得在使用的时候保持一致就可以了；"content://com.example.MyContentProvider/users"。
- D：如果 URI 中包含表示需要获取的记录的 ID，则就返回该 id 对应的数据；如果没有 ID，就表示返回全部；"content://com.example.MyContentProvider/users/100" 100 表示数据 id。

2. 系统通讯录核心操作代码

本项目的课程学习我们将以操作系统通讯录为例，进行学习。下面展示了一些核心代码。

1) 读取联系人信息

```
lvContent = (ListView) findViewById(R.id.lv_content);
Cursor cursor = getContentResolver().query(
ContactsContract.Contacts.CONTENT_URI, null, null, null, null);
SimpleCursorAdapter adapter = new SimpleCursorAdapter(this,
android.R.layout.simple_list_item_1, cursor,
new String[] { ContactsContract.Contacts.DISPLAY_NAME },
new int[] { android.R.id.text1 });
lvContent.setAdapter(adapter);
```

读取系统联系人需要 ContentResolver.query()方法，该方法定义如下：

```
public final Cursor query(Uri uri, String[] projection, String selection, String[] selectionArgs, String sortOrder) {
    return query(uri, projection, selection, selectionArgs, sortOrder, null);
}
```

query 方法返回一个 Cursor 对象，这个对象和查询 SQLite 数据库返回的 Cursor 对象一样，可以直接访问 Cursor 对象中的数据，也可以将其和 CursorAdapt 一起使用，其中 5 个参数含义如下：

- uri：查询地址，如 Uri.parse("content://com.android.contacts/contacts")；或者 ContactsContract.Contacts.CONTENT_URI
- projection：需要查询的字段，类似 SQL 语言中 select 和 from 之间的部分，例如：name，salary
- selection：查询条件，类似 where 后面的语句，例如：name=? andsalary=?

- selectionArgs：占位符，问号表示的部分，例如：new String[]{"bill"，"1200"}
- sortQuery：需要排序的字段，order by 后面的部分，例如：name，salary desc

2）通过名字查找联系人

Cursor cur = getContentResolver().query(ContactsContract.Contacts.CONTENT_URI, null, ContactsContract.Contacts.DISPLAY_NAME+ " = '" + "刘德华"+"'", null, null);

3）查询 ID

Uri contactsUri = Uri.parse("content://com.android.contacts/contacts");
Cursor contentBaseCursor = getContentResolver().query(contactsUri,
new String[] { "_id", "display_name", "sort_key" }, null, null, "sort_key");
contentBaseCursor.moveToFirst();
System.out.println("count-->" + contentBaseCursor.getCount());
for (int i = 0; i < contentBaseCursor.getCount(); i++) {
    String str = contentBaseCursor.getString(0);
    System.out.println(str);
    contentBaseCursor.moveToNext();
}

4）查询联系人

Uri uri2 = ContactsContract.Contacts.CONTENT_URI;
Cursor cursor = getContentResolver().query(uri2, null, null, null, null);
cursor.moveToFirst();
while (cursor.moveToNext()) {
System.out.println(cursor.getString(cursor.getColumnIndex(ContactsContract.Contacts.DISPLAY_NAME)));
}

5）查询联系人电话

Uri uri = ContactsContract.CommonDataKinds.Phone.CONTENT_URI;
Cursor c = getContentResolver().query(uri,
new String[] { ContactsContract.PhoneLookup.DISPLAY_NAME,
ContactsContract.CommonDataKinds.Phone.NUMBER }, null, null, null);
c.moveToFirst();
System.out.println("c==" + c.getCount());
while (c.moveToNext()) {
System.out.println("name-->" + c.getString(0) + ", num-->"+ c.getString(1));
}

6）调用系统方法查询联系人

Intent intent = new Intent();
intent.setAction("com.android.contacts.action.FILTER_CONTACTS");
intent.setType("vnd.android.cursor.dir/contact");
intent.addCategory("android.intent.category.DEFAULT");
startActivity(intent);

7）传递参数查找联系人

Intent intent = new Intent();
intent.putExtra(SearchManager.QUERY, "李志");
intent.setAction(Intent.ACTION_SEARCH);
intent.setPackage("com.android.contacts");
startActivity(intent);

8）联系人保存位置

联系人数据库保存在 data/data/com.android.providers.contacts/database/contacts2.db 中。其中主要的表有 raw_contacts（保存联系人 id），mimetypes（保存数据类型），data（保存详细数据）。

- raw_contacts：存放联系人的 ID，_id 属性为主键，声明为 autoincrement，即不需要手动设置，其他属性也不需要手动设置就有默认值；display_name 属性为姓名。
- mimetypes：存放数据的类型，比如"vnd.android.cursor.item/name"表示姓名类型的数据，"vnd.android.cursor.item/phone_v2"表示电话类型的数据。
- data：存放具体的数据；raw_contact_id 属性用来连接 raw_contacts 表，每条记录表示一个具体数据；我们主要的数据（email、phone 等)都存放在 data 表。

三、项目实施过程

下面我们通过一个简单的通讯录项目来进行本章项目的学习。

1. 工程创建

在 eclipse 中新建一个名为 AndroidCode14 的 Android 项目，新建 com.xdxy.contentprovider 包，包下新建 ContactBean、ItemsAdapter 和 MainActivity 三个 Java 文件，如图 18-2 所示。

- ContactBean：联系人实体类。
- ItemAdapter：联系列表的子项布局适配器。

项目十八 ContentProvider 实现数据共享

➢ MainActivity：主页面活动类。

2. XML 布局文件的开发

程序的 XML 布局文件包括 activity_main、contact_item、insert_layout 和 update_layout。主界面 activity_main 显示程序的主界面，如图 18-3 所示。顶部 1 个 EditText 输入查询的关键字（姓名或电话），中部 1 个 ListView 显示查询结果，下部 3 个 Button，分别表示查询，插入，退出。

图 18-2　工程目录结构图

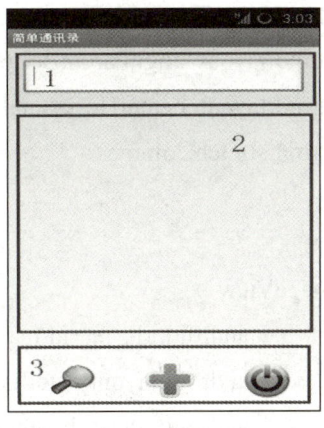

图 18-3　通讯录主界面

1）activity_main

（1）以下代码显示标号 1 部分的控件，由一个 EditText 输入框组成，功能是输入查询内容。

```
<EditText
        android:id="@+id/main_et_input"
        android:layout_width="fill_parent"
        android:layout_height="wrap_content"
        android:layout_alignParentTop="true"
        android:inputType="text" />
```

（2）以下代码显示标号 2 部分，由一个 ListView 控件组成，功能是显示可下拉的文本框。

```
<ListView
        android:id="@+id/lv_content"
        android:layout_width="fill_parent"
        android:layout_height="fill_parent"
```

```
            android:layout_above="@id/btn_query"
            android:layout_below="@id/main_et_input" />
```

(3)以下代码显示标号 3 部分，在 TabLayout 布局中放置三个 ImageView，功能是设定查询，添加，退出按钮。

```xml
<TableLayout
        android:id="@+id/btn_query"
        android:layout_width="wrap_content"
        android:layout_height="wrap_content"
        android:layout_alignParentBottom="true"
        android:layout_centerHorizontal="true"
        android:stretchColumns="*" >
<TableRow>
<!-- 查询 -->
        <ImageView
                android:id="@+id/main_iv_query"
                android:layout_width="100dp"
                android:layout_height="50dp"
                android:contentDescription="@id/main_iv_query"
                android:src="@drawable/find_icon" />
<!-- 添加 -->
        <ImageView
                android:id="@+id/main_iv_add"
                android:layout_width="100dp"
                android:layout_height="50dp"
                android:contentDescription="@id/main_iv_add"
                android:src="@drawable/add_icon" />
<!-- 退出 -->
        <ImageView
                android:id="@+id/main_iv_exit"
                android:layout_width="100dp"
                android:layout_height="50dp"
                android:contentDescription="@id/main_iv_add"
                android:src="@drawable/exit_icon" />
```

```xml
</TableRow>
</TableLayout>
```

2) contact_item_layout

contact_item_layout 的作用是显示每条记录的显示,由一个 TableLayout 布局组成,里面包含了两个 TextView,分别显示姓名和电话。

```xml
<?xml version="1.0" encoding="utf-8"?>
<RelativeLayout xmlns:android="http://schemas.android.com/apk/res/android"
    android:layout_width="match_parent"
    android:layout_height="match_parent" >
<!-- 通讯录 item -->
<TableLayout
        android:layout_width="wrap_content"
        android:layout_height="wrap_content"
        android:layout_centerHorizontal="true" >
<TableRow>
        <TextView
                android:id="@+id/item_tv_name"
                android:layout_width="150dp"
                android:layout_height="wrap_content"
                android:layout_gravity="center|left"
                android:layout_margin="10dp" />
        <TextView
                android:id="@+id/item_tv_phone"
                android:layout_width="150dp"
                android:layout_height="wrap_content"
                android:layout_gravity="center|left"
                android:layout_margin="10dp" />
</TableRow>
</TableLayout>
</RelativeLayout>
```

3) insert_layout

insert_layout 是添加新联系人的弹出窗口布局,由一个 TableLayout 布局包裹两个 TextView 和两个 EditText 构成,分别显示添加的联系人姓名和电话。

```xml
<?xml version="1.0" encoding="utf-8"?>
<RelativeLayout xmlns:android="http://schemas.android.com/apk/res/android"
    android:layout_width="fill_parent"
    android:layout_height="fill_parent" >
    <TableLayout
        android:layout_width="wrap_content"
        android:layout_height="wrap_content"
        android:layout_centerHorizontal="true"
        android:layout_centerVertical="true" >
<TableRow>
        <TextView
            android:layout_width="wrap_content"
            android:layout_height="wrap_content"
            android:layout_margin="5dp"
            android:text="@string/str_name" />
        <EditText
            android:id="@+id/ins_ly_et_name"
            android:layout_width="100dp"
            android:layout_margin="5dp"
            android:inputType="text" />
</TableRow>
<TableRow>
        <TextView
            android:layout_width="wrap_content"
            android:layout_height="wrap_content"
            android:layout_margin="5dp"
            android:text="@string/str_number" />
        <EditText
            android:id="@+id/ins_ly_et_number"
            android:layout_width="170dp"
            android:layout_margin="5dp"
            android:inputType="number"
            android:maxLength="14" />
```

```
</TableRow>
</TableLayout>
</RelativeLayout>
```

4) update_layout

update_layout 是修改联系人电话的弹出窗口布局，由 TableLayout 布局包裹一个 TextView 和一个 EditText 组成。

```
<?xml version="1.0" encoding="utf-8"?>
<RelativeLayout xmlns:android="http://schemas.android.com/apk/res/android"
    android:layout_width="fill_parent"
    android:layout_height="fill_parent" >
    <TableLayout
        android:layout_width="wrap_content"
        android:layout_height="wrap_content"
        android:layout_centerHorizontal="true"
        android:layout_centerVertical="true" >
<TableRow>
        <TextView
                android:layout_width="wrap_content"
                android:layout_height="wrap_content"
                android:layout_margin="5dp"
                android:text="@string/str_number" />
        <EditText
                android:id="@+id/upd_ly_et_number"
                android:layout_width="170dp"
                android:layout_margin="5dp"
                android:inputType="number"
                android:maxLength="14" />
</TableRow>
</TableLayout>
</RelativeLayout>
```

3. Java 文件的开发

程序分为 3 个 class 文件：ContactBean 存储联系人的属性字段，ItemsAdapter 是自定义适配器,MainActivity 是主函数。

(1) ContactBean 类,本类存储联系人的编号、姓名、电话、public 静态常量存储关键字和 get、set 方法使得其他类能够在外部访问。

```java
public class ContactBean {
    // 常量
    public static String CONTACT_ID = "_id";
    public static String CONTACT_NAME = "name";
    public static String CONTACT_PHONE = "phone";
    private int _id;
    private String name;
    private String phone;

    public ContactBean() {
        // TODO Auto-generated constructor stub
    }

    public ContactBean(int id, String name, String phone) {
        this._id = id;
        this.name = name;
        this.phone = phone;

    }

    public int get_id() {
        return _id;
    }

    public void set_id(int _id) {
        this._id = _id;
    }

    public String getName() {
        return name;
    }
```

```java
public void setName(String name) {
    this.name = name;
}

public String getPhone() {
    return phone;
}

public void setPhone(String phone) {
    this.phone = phone;
}
}
```

（2）ItemsAdapter 类，本类关键方法 getView 获取 litView 里显示的 item 字段信息，并用重载 convertView 的方法优化数据显示。

```java
public class ItemsAdapter extends BaseAdapter {
    LayoutInflater inflater;
    List<Map<String, Object>> listItem;
    String index[];
    Context context;
    public ItemsAdapter(List<Map<String, Object>> list, String index[],Context context) {
        this.index = index;
        this.context = context;
        inflater = LayoutInflater.from(context);
        this.listItem = list;
    }
    @Override
    public int getCount() {
        // TODO Auto-generated method stub
        return listItem.size();
    }
    @Override
    public Object getItem(int position) {
        // TODO Auto-generated method stub
        return listItem.get(position);
    }
```

```java
        @Override
        public long getItemId(int position) {
            // TODO Auto-generated method stub
            return position;
        }
        @Override
        public View getView(int position, View convertView, ViewGroup parent) {
            ViewHolder holder = null;
            if (convertView == null) {
                holder = new ViewHolder();
                // 获取字符串
                String strName = listItem.get(position).get(index[0]) + "";
                String strPhone = listItem.get(position).get(index[1]) + "";
                // 获取文件视图
                convertView = inflater.inflate(R.layout.contact_item_layout, null);
                holder.tvName = (TextView) convertView
                        .findViewById(R.id.item_tv_name);
                holder.tvPhone = (TextView) convertView
                        .findViewById(R.id.item_tv_phone);
                // 设置字符串内容
                holder.tvName.setText(strName);
                holder.tvPhone.setText(strPhone);
                // 加入 convertView
                convertView.setTag(holder);
            } else {
                holder = (ViewHolder) convertView.getTag();
            }
            return convertView;
        }

        public final class ViewHolder {
            public TextView tvId, tvName, tvPhone;
        }
    }
```

（3）MainActivity 类，本类通过 onCreate()方法调用 initView()初始化控件并用 findViewById()建立与 XML 文件的关联，initeListener()设置监听器使得单击 button 按钮是

调用对应的处理方法。

```java
/**
 * @description 初始化显示数据
 */
private void initView() {
    context = MainActivity.this;
    listViewItem = (ListView) findViewById(R.id.lv_content);
    contactList = new ArrayList<ContactBean>();
    etInput = (EditText) findViewById(R.id.main_et_input);
    // 点击删除
    imgQuery = (ImageView) findViewById(R.id.main_iv_query);
    imgAdd = (ImageView) findViewById(R.id.main_iv_add);
    imgExit = (ImageView) findViewById(R.id.main_iv_exit);
}
private void initListener() {
    // 添加联系人
    imgAdd.setOnClickListener(new OnClickListener() {

        @Override
        public void onClick(View v) {
            // 弹出对话框显示插入界面
            // 载入 xml 文件的布局
            LayoutInflater lf = (LayoutInflater) MainActivity.this
                    .getSystemService(Context.LAYOUT_INFLATER_SERVICE);
            View vg = (View) lf.inflate(R.layout.insert_layout, null);
            vg.setBackgroundColor(Color.WHITE);
            final EditText etName = (EditText) vg
                    .findViewById(R.id.ins_ly_et_name);
            final EditText etNumber = (EditText) vg
                    .findViewById(R.id.ins_ly_et_number);

            new AlertDialog.Builder(MainActivity.this)
                    .setView(vg)
                    .setPositiveButton("确定",
                            new DialogInterface.OnClickListener() {
```

```java
                                @Override
                                public void onClick(DialogInterface dialog,
                                        int which) {
                                    strName = etName.getText().toString();
                                    strNumber = etNumber.getText()
                                            .toString();
                                    System.out.println("str--->" + strName
                                            + "num:" + strNumber);
                                    insertValue(strName, strNumber);
                                }
                            }).setNegativeButton("取消", null).show();

            }
        });

        // 查找数据
        imgQuery.setOnClickListener(new OnClickListener() {

            @Override
            public void onClick(View v) {

                // 首拼，全拼查找联系人
                contactList = new ArrayList<ContactBean>(getQuery(etInput
                        .getText().toString()));
                if (contactList.size() == 0) {
                    // 清空所有显示数据
                    showListView();
                    Toast.makeText(context, "无此数据！", Toast.LENGTH_SHORT).show();
                } else {
                    showListView();
                }
            }
        });
        // 退出
        imgExit.setOnClickListener(new OnClickListener() {
```

```java
            @Override
            public void onClick(View v) {
                // TODO Auto-generated method stub
                finish();
            }
        });
        // 删除和修改
        listViewItem.setOnItemClickListener(new OnItemClickListener() {

            @SuppressWarnings("unchecked")
            @Override
            public void onItemClick(AdapterView<?> parent, View view,
                    int position, long id) {
                map = (Map<String, Object>) myAdapter.getItem(position);
                nameQuery = map.get(ContactBean.CONTACT_NAME).toString();
                idQuery = map.get("_id").toString();
                System.out.println("选择的 name-->" + nameQuery);

                new AlertDialog.Builder(context)
                        .setTitle("对本条记录如何操作？")
                        .setPositiveButton("删除",
                                new DialogInterface.OnClickListener() {
                                    public void onClick(DialogInterface dialog,
                                            int whichButton) {
                                        // 删除记录
                                        deleteContact(nameQuery);
                                    }
                                })
                        .setNeutralButton("修改",
                                new DialogInterface.OnClickListener() {

                                    @Override
                                    public void onClick(DialogInterface dialog,
                                            int which) {
                                        doUpdateData();
```

```
                            }
                        })
                .setNegativeButton("取消",
                        new DialogInterface.OnClickListener() {

                            @Override
                            public void onClick(DialogInterface dialog,
                                    int which) {

                            }
                        }).show();

            }
        });

    }
```

deleteContact()方法，在查询出来的数据项上右击调用contentProvider的delete方法删除条目。

```
private void deleteContact(String nameQuery) {
    try {
        // String name = "aaa";
        // 根据姓名求 id
        Uri uri = Uri.parse("content://com.android.contacts/raw_contacts");
        ContentResolver resolver = context.getContentResolver();
        Cursor cursor = resolver.query(uri, new String[] { Data._ID },
                "display_name=?", new String[] { nameQuery }, null);
        if (cursor.moveToFirst()) {
            int id = cursor.getInt(0);
            // 根据 id 删除 data 中的相应数据
            resolver.delete(uri, "display_name=?",new String[] { nameQuery });
            uri = Uri.parse("content://com.android.contacts/data");
            resolver.delete(uri, "raw_contact_id=?",new String[] { id + "" });
        }
        // 重新查询，刷新显
        freshData(etInput.getText().toString());
```

```
            Toast.makeText(context, "删除成功", Toast.LENGTH_SHORT).show();
        } catch (Exception e) {
            Toast.makeText(context, "删除失败, 请重试, Toast.LENGTH_SHORT).show();
            e.printStackTrace();
        }
    }
```

updateData()方法,点查询出来的图书记录上右击可以选择修改图书信息,调用contentProvider的update方法修改图书信息。

```
private void updateData(String id, String name, String number) {
ArrayList<ContentProviderOperation> ops = new ArrayList<ContentProviderOperation>();
    try {
        // // 修改电话号码
        ops.add(ContentProviderOperation.newUpdate(Data.CONTENT_URI)
                .withSelection("_id =?", new String[] { id })
                .withValue(Phone.NUMBER, number).build());
        getContentResolver().applyBatch(ContactsContract.AUTHORITY, ops);
        freshData(etInput.getText().toString());
        Toast.makeText(context, "修改成功", Toast.LENGTH_SHORT).show();
    } catch (Exception e) {
        Toast.makeText(context, "修改失败", Toast.LENGTH_SHORT).show();
        e.printStackTrace();
    }
}
```

getQuery()方法,在输入框输入字符串调用contentProvider的query方法进行模糊查询,可以根据名字和姓名查询。

```
private ArrayList<ContactBean> getQuery(String key) {
    ArrayList<ContactBean> list = new ArrayList<ContactBean>();
Uri uri = Uri.
withAppendedPath(ContactsContract.CommonDataKinds.Phone.
                                    CONTENT_FILTER_URI,Uri.encode(key));
    Cursor cursor = getContentResolver().query(uri,
    new String[] { ContactsContract.CommonDataKinds.Phone._ID, // "_id"
ContactsContract.CommonDataKinds.Phone.DISPLAY_NAME, // "display_name"
ContactsContract.CommonDataKinds.Phone.NUMBER }, null, // "data1"
    null, null);
System.out.println();
```

```
            System.out.println("total=========>" + cursor.getCount());
            cursor.moveToFirst();
            for (int i = 0; i < cursor.getCount(); i++) {
                String id = cursor.getString(0);
                String name = cursor.getString(1);
                String number = cursor.getString(2);
                System.out.println("id:" + id + ".name:" + name + ",number:"+ number);
                // 查到的结果加入 list
                ContactBean contact = new ContactBean(Integer.parseInt(id), name, number);
                list.add(contact);
                cursor.moveToNext();
            }
            cursor.close();
            return list;
        }
```

4. 项目验证

（1）将 Android 开发终端与 PC 机相连，并运行程序如图 18-4 所示。

（2）在输入框输入联系人姓名或电话号码可以后单击 🔍 查询出联系人信息，如图 18-5 所示。联系人支持全拼、首字查找，如图 18-6 所示。支持电话查询，如图 18-7 所示。

图 18-4　查询运行主界面

图 18-5　全拼查询

图 18-6　首拼查询　　　　　　　　　图 18-7　电话查询

（3）在主界面单击添加按钮可以添加联系人，如图 18-8 和图 18-9 所示。

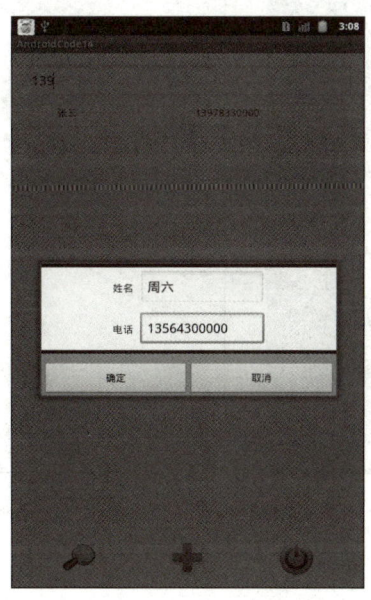

图 18-8　添加联系人　　　　　　　　图 18-9　联系人添加成功

（4）在查询出的结果列表中单击单条记录，在弹出的对话框可以选择【删除】或者【修改】联系人信息，如图 18-10 所示。删除联系人的结果如图 18-11 所示，修改联系人效果如图 18-12、图 18-13 所示和图 18-14 所示。

图 18-10　删除联系人

图 18-11　联系人删除成功

图 18-12　修改联系人电话

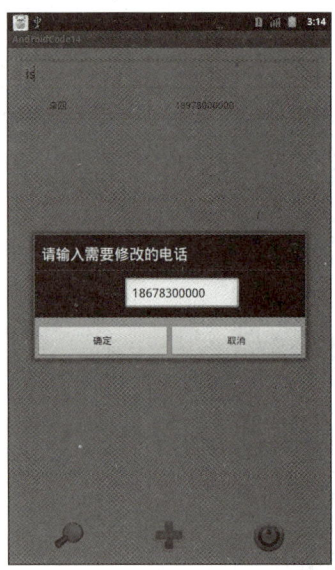
图 18-13　输入新的电话号码

图 18-14　联系人电话修改成功

四、项目思考与扩展

参照正文实验修改程序，在添加联系人时加入 Email。并在查询联系人时在列表中显示 Email。

扩展篇

项目十九 位置服务与百度地图实现地图定位

【本章导读】

通过之前的学习,我们已经学习到了许多的 Android 知识。本小节,我们将学习只有移动设备上才能实现的技术,基于位置的服务(Location Based Service)。通过地理定位的技术就可以随时得知自己所在的位置,围绕这一点开发出很多有意思的应用。

一、项目要求

1. 熟悉 Android 应用程序开发流程。
2. 掌握 LocationManager 的使用方法。
3. 掌握百度地图密钥申请与地图使用方法。

二、项目相关知识

1. 基于位置的服务简介

基于位置的服务简称 LBS,这个技术随着移动互联网的兴起,在最近的几年里十分火爆。其实它本身并不是什么时髦的技术,主要的工作原理就是利用无线电通信网络或 GPS 等定位方式来确定出移动设备所在的位置,其实这种定位技术早在很多年前就已经出现了。

那为什么 LBS 技术直到最近几年才开始流行呢？这主要是因为，在过去移动设备的功能极其有限，即使定位到了设备所在的位置，也就仅仅只是定位到了而已，我们并不能在位置的基础上进行一些其他的操作。而现在就大大不同了，有了 Android 系统作为载体，我们可以利用定位出的位置进行许多丰富多彩的操作。

比如说天气预报程序可以根据用户所在的位置自动选择城市，发微博的时候我们可以向朋友们晒一下自己在哪里，不认识路的时候随时打开地图就可以查询路线，等等。归根结底，其实基于位置的服务所围绕的核心就是要确定出自己所在的位置，这在 Android 中并不困难，主要借助 LocatinManager 这个类就可以实现了。

下面我们首先学习一下 LocationManager 的基本用法，然后再通过一个例子来尝试获取一下自己当前的位置。

> 另外需要注意，本项目中所写的程序建议大家都在手机上运行，DDMS 虽然也提供了在模拟器中模拟地理位置的功能，但在手机上得到真实的位置数据，你的感受会更加深刻。

2. LocationManager 的基本用法

毫无疑问，要想使用 LocationManager 就必须要先获取到它的实例，我们可以调用 Context 的 getSystemService()方法获取到。getSystemService()方法接收一个字符串参数用于确定获取系统的哪个服务，这里传入 Context.LOCATION_SERVICE 即可。因此，获取 LocationManager 的实例就可以写成：

> LocationManagerlocationManager=(LocationManager)getSystemService(
> Context.LOCATION_SERVICE);

接着我们需要选择一个位置提供器来确定设备的当前位置。Android 中一般有三种位置提供器可供选择，GPS_PROVIDER、NETWORK_PROVIDER 和 PASSIVE_PROVIDER。其中前两种用的比较多，分别表示使用 GPS 定位和使用网络定位。

这两种定位方式各有特点，GPS 定位的精准度比较高，但是非常耗电；而网络定位的精准度稍差，但耗电量比较少。我们应该根据自己的实际情况来选择使用哪一种位置提供器，当位置精度要求非常高的时候，最好使用 GPS_PROVIDER，而一般情况下，使用 NETWORK_PROVIDER 会更加划算。

需要注意的是，定位功能必须要由用户主动去启用才行，不然任何应用程序都无法获取到手机当前的位置信息。进入手机的【设置】→【定位服务】，其中第一个选项表示允许使用网络的方式来对手机进行定位,第二个选项表示允许使用 GPS 的方式来对手机进行定位。如图 19-1 所示。

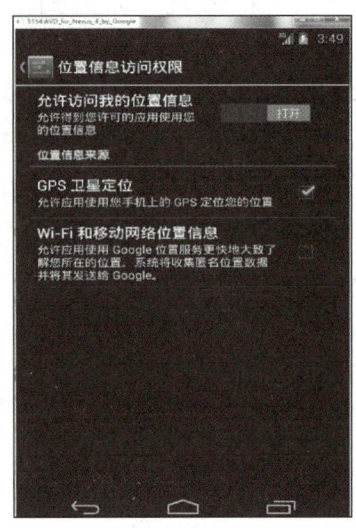

图 19-1　GPS 定位

这些选项只是表明你已经同意让应用程序来对你的手机进行定位了,但只有当定位操作真正开始的时候才会影响到手机的电量。下面我们就来看一看,如何才能真正地开始定位操作。

将选择好的位置提供器传入到 getLastKnownLocation()方法中,就可以得到一个 Location 对象,如下所示:

String provider = LocationManager.NETWORK_PROVIDER;
Location location = locationManager.getLastKnownLocation(provider);

这个 Location 对象中包含了经度、纬度、海拔等一系列的位置信息,然后从中取出我们所关心那部分数据即可。

如果有些时候你想让定位的精度尽量高一些,但又不确定 GPS 定位的功能是否已经启用,这个时候就可以先判断一下有哪些位置提供器可用,如下所示:

List<String> providerList = locationManager.getProviders(true);

可以看到,getProviders()方法接收一个布尔型参数,传入 true 就表示只有启用的位置提供器才会被返回。之后再从 providerList 中判断是否包含 GPS 定位的功能就行了。

另外,调用 getLastKnownLocation()方法虽然可以获取到设备当前的位置信息,但是用户是完全有可能带着设备随时移动的,那么我们怎样才能在设备位置发生改变的时候获取到最新的位置信息呢?

项目十九 位置服务与百度地图实现地图定位

不用担心，LocatinManager 还提供了一个 requestLocationUpdates()方法，只要传入一个 LocationListener 的实例，并简单配置几个参数就可以实现上述功能了，写法如下：

```
locationManager.requestLocationUpdates(provider, 5000, 10,   new LocationListener() {
    @Override
    public void onStatusChanged(String provider, int status, Bundle extras) {
}
@Override
public void onProviderEnabled(String provider) {
}

@Override
public void onProviderDisabled(String provider) {
}
@Override
public void onLocationChanged(Location location) {
}
});
```

这里 requestLocationUpdates()方法接收四个参数：第一个参数是位置提供器的类型；第二个参数是监听位置变化的时间间隔，以毫秒为单位；第三个参数是监听位置变化的距离间隔，以米为单位；第四个单位则是 LocationListener 监听器。

这样的话，LocationManager 每隔 5 秒会检测一下位置的变化情况，当移动距离超过 10 米的时候，就会调用 LocationListener 的 onLocationChanged()方法，并把新的位置信息作为参数传入。

3．获取 GPS 定位信息的步骤

（1）获取系统的 LocationManager 对象。

（2）使用 LocationManager，通过指定 LocationProvider 来获取定位信息，定位信息由 Location 对象来表示。

（3）从 Location 对象中获取定位信息。

4．使用 MapView 显示定位的过程

（1）获取 MapView 对应的 MapController 对象。
（2）根据程序获取的经度、纬度创建 GeoPoint 对象。

（3）调用 MapView 所关联的 MapController 对象的 animateTo(GeoPoint point)方法定位到指定位置。

（4）获取 MapView 上屏幕坐标与经纬度坐标间的投影关系。

（5）调用 Projection 的 toPixels 方法把经纬度坐标转换为屏幕坐标。

（6）调用 Canvas 的 drawBitmap 方法在屏幕上绘制图片。

三、项目实施过程

下面我们来做一个实例，利用百度地图实现 Andorid 手机的定位。效果如图 19-2 所示。

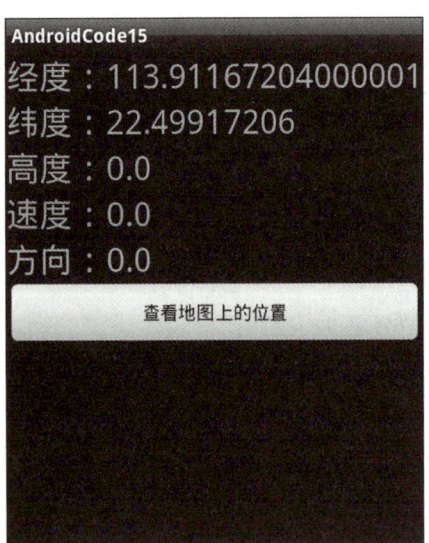

图 19-2　项目效果图

1. 工程创建

（1）获取百度地图密钥，然后输入你的百度账号登录，如果没有账号就必须申请账号，如图 19-3 所示。

项目十九　位置服务与百度地图实现地图定位

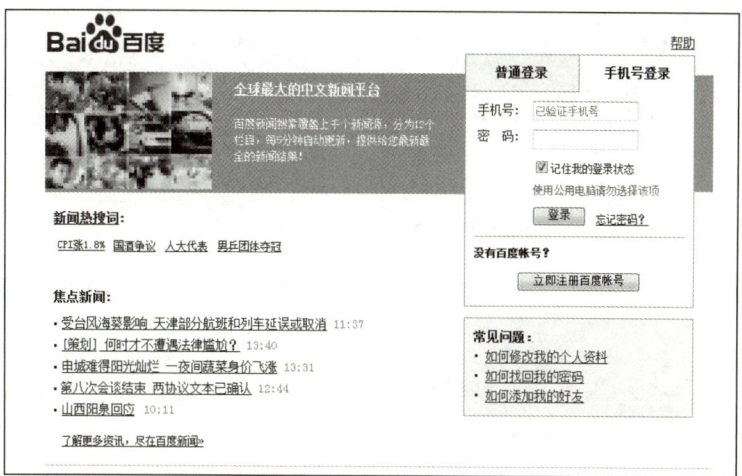

图 19-3　登录百度帐户

（2）登录成功后输入勾选同意条款、输入应用名称和验证码单击生成 API 密钥按钮生成密钥，如图 19-4 所示。

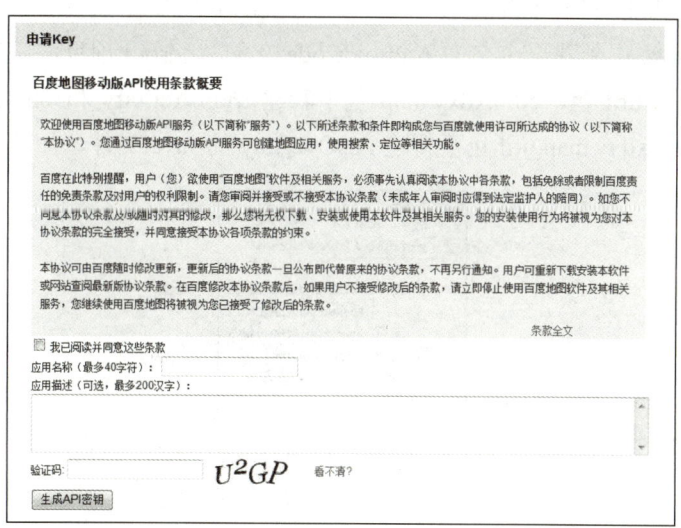

图 19-4　生成百度地图密钥

密钥生成，如图 19-5 所示。

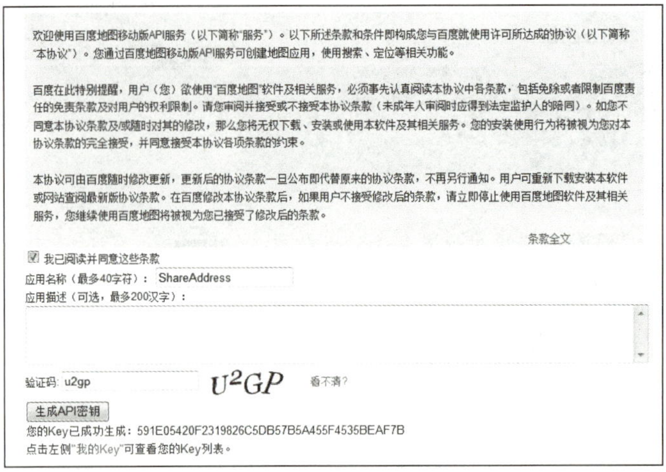

图 19-5 百度地图密钥

至此，我们就获得了所有开发阶段的应用程序通用的 Map API 密钥。不过需要注意的是，在应用程序发布时，需要根据为应用程序签名的密钥重新生成 Map API 密钥，并在程序中修改引用到 Map API 密钥的地方。

（3）在 eclipse 中新建一个名为 AndroidCode19 的 Android 项目，新建 com.xdxy.map 包和 com.xdxy.map.util 包，com.xdxy.map 包下新建 MainActivity 和 ShowMapActivity，2 个 Java 文件，com.xdxy.map.util 包下新建 PosOverLay 的 Java 文件，如图 19-6 所示。

图 19-6 工程目录

2．XML 布局文件的开发

（1）地理位置共享程序的 XML 布局文件（main.xml），包含 2 个显示界面包括登主界面、地图界面。主界面使用 TextView 来显示 GPS 获取到的信息，程序如下所示：

```xml
<?xml version="1.0" encoding="utf-8"?>
<LinearLayout xmlns:android="http://schemas.android.com/apk/res/android"
    android:orientation="vertical"
    android:layout_width="fill_parent"
    android:layout_height="fill_parent">
    <TableLayout
        android:layout_width="fill_parent"
        android:layout_height="wrap_content">
    <TableRow>
    <TextView
        android:layout_width="fill_parent"
        android:layout_height="fill_parent"
        android:text="经度："
        android:textSize="25dip"
        android:gravity="center"/>
    <TextView
        android:id="@+id/jingdu"
        android:layout_width="fill_parent"
        android:layout_height="fill_parent"
        android:textSize="25dip"/>
    </TableRow>
    <TableRow>
    <TextView
        android:layout_width="fill_parent"
        android:layout_height="fill_parent"
        android:text="纬度："
        android:gravity="center"
        android:textSize="25dip"/>
    <TextView
        android:id="@+id/weidu"
        android:layout_width="fill_parent"
        android:layout_height="fill_parent"
        android:textSize="25dip"/>
```

```xml
        </TableRow>
        <TableRow>
        <TextView
            android:layout_width="fill_parent"
            android:layout_height="fill_parent"
            android:text="高度："
android:gravity="center"
android:textSize="25dip"/>
        <TextView
            android:id="@+id/gaodu"
            android:layout_width="fill_parent"
            android:layout_height="fill_parent"
            android:textSize="25dip"/>
        </TableRow>
        <TableRow>
        <TextView
            android:layout_width="fill_parent"
            android:layout_height="fill_parent"
            android:text="速度："
          android:gravity="center"
          android:textSize="25dip"/>
        <TextView
            android:id="@+id/sudu"
            android:layout_width="fill_parent"
            android:layout_height="fill_parent"
            android:textSize="25dip"/>
        </TableRow>
        <TableRow>
        <TextView
            android:layout_width="fill_parent"
            android:layout_height="fill_parent"
            android:text="方向："
            android:gravity="center"
```

```
            android:textSize="25dip"/>
        <TextView
            android:id="@+id/fangxiang"
            android:layout_width="fill_parent"
            android:layout_height="fill_parent"
            android:textSize="25dip"/>
    </TableRow>
    </TableLayout>
    <Button
        android:id="@+id/showMapAddress"
        android:layout_width="fill_parent"
        android:layout_height="wrap_content"
        android:text="查看地图上的位置"/>
</LinearLayout>
```

地图显示界面的 XML 文件,是由 RadioGroup 中的 2 个 RadioButton 组成的,使用 MapView 来显示百度地图内容以及绘制位置图标,具体程序如下:

```
<?xml version="1.0" encoding="utf-8"?>
<LinearLayout
    xmlns:android="http://schemas.android.com/apk/res/android"
    android:layout_width="fill_parent"
    android:layout_height="fill_parent"
    android:orientation="vertical">
<RadioGroup
    android:id="@+id/rg"
    android:orientation="horizontal"
    android:layout_width="wrap_content"
    android:layout_height="wrap_content">
<RadioButton
    android:id="@+id/generalMap"
    android:layout_width="wrap_content"
    android:layout_height="wrap_content"
    android:checked="true"
    android:text="普通地图"/>
```

```xml
<RadioButton
    android:id="@+id/SatelliteMap"
    android:layout_width="wrap_content"
    android:layout_height="wrap_content"
    android:text="卫星地图"/>
</RadioGroup>
<com.baidu.mapapi.MapView android:id="@+id/bmapsView"
    android:layout_width="fill_parent"
    android:layout_height="fill_parent"
    android:clickable="true" />
</LinearLayout>
```

3. Java 文件的开发

（1）MainActivity 类主要用于显示 GPS 定位信息，以及实现共享位置信息、在地图上显示位置信息等操作。首先重写 onCreate()方法，在方法中首先设置视图初试化各类组件，为按钮组件添加监听事件，然后获取 LocationManager 对象，通过 LocationManager 对象的 getLastKnownLocation()方法获取 Loaction 对象，最后使用 Location 对象的 getLongitude()、getLatitude()等获得当前位置的经度、纬度等信息，调用 updataview()显示、更新定位信息。关键部分的程序如下：

```java
public void onCreate(Bundle savedInstanceState) {
    super.onCreate(savedInstanceState);
    setContentView(R.layout.main);
    // 实例化组件
    jingdu = (TextView) findViewById(R.id.jingdu);
    weidu = (TextView) findViewById(R.id.weidu);
    gaodu = (TextView) findViewById(R.id.gaodu);
    sudu = (TextView) findViewById(R.id.sudu);
    fangxiang = (TextView) findViewById(R.id.fangxiang);
    showMapAddress = (Button) findViewById(R.id.showMapAddress);
    // 设置监听事件
    showMapAddress.setOnClickListener(new OnButtonClick());
    handler = new Handler() {
        @Override
```

```java
            public void handleMessage(Message msg) {
                // TODO Auto-generated method stub
                super.handleMessage(msg);
                if (msg != null) {
                    Bundle bundle = msg.getData();
                    jingdu.setText(bundle.getString("jingdu"));
                    weidu.setText(bundle.getString("weidu"));
                    gaodu.setText(bundle.getString("gaodu"));
                    sudu.setText(bundle.getString("sudu"));
                    fangxiang.setText(bundle.getString("fangxiang"));
                }
            }
        };
    }
    protected void onStart() {
        // TODO Auto-generated method stub
    super.onStart();

    // 创建 LocationManager 对象
    loc = (LocationManager) getSystemService(Context.LOCATION_SERVICE);
    // 创建 Location 对象从 GPS 中获取定位信息
    locaction = loc.getLastKnownLocation(LocationManager.GPS_PROVIDER);
    updateview(locaction);
    // 设置 3 秒获取一次 GPS 定位信息
    loc.requestLocationUpdates(LocationManager.GPS_PROVIDER,
            3000, 8,new LocationListener() {
                @Override
                public void onStatusChanged(String provider,
                        int status,Bundle extras) {
                }
                @Override
                public void onProviderEnabled(String provider){
                    // TODO Auto-generated method stub
```

```
            // 更新位置信息
            updateview(loc.getLastKnownLocation(provider));
        }
        @Override
        public void onProviderDisabled(String provider) {
            // TODO Auto-generated method stub
            loc_net();
        }
        @Override
        public void onLocationChanged(Location location) {
            // TODO Auto-generated method stub
            // 更新位置信息
                    updateview(locaction);
                }
    });
}
```

（2）ShowMapActivity 类的主要功能是把 GPS 获得的位置信息以小泡的形式显示在 Google 地图上。首先 ShowMapActivity 必须继承 MapActivity 类，然后获取 Mapview 对象并接收 MianActivity 传递过来的定位数据，调用 upMapview()方法将位置信息显示到地图上，再通过实例化 PosOverLay 类来实现在地图上绘制小泡。关键部分的程序如下：

```
    protected void onCreate(Bundle arg0) {
        // TODO Auto-generated method stub
        super.onCreate(arg0);
        setContentView(R.layout.mlayout);
        // 根据组件 ID 获取组件对象
        rg = (RadioGroup) findViewById(R.id.rg);
        generalMap = (RadioButton) findViewById(R.id.generalMap);
        SatelliteMap = (RadioButton) findViewById(R.id.SatelliteMap);

        mBMapMan = new BMapManager(getApplication());
        mBMapMan.init("7994E9433E9A31FA26B73BA6F7596D43DAF5A32D", null);
        super.initMapActivity(mBMapMan);
        mMapView = (MapView) findViewById(R.id.bmapsView);
```

```java
        mMapView.setBuiltInZoomControls(true);
    }
    @Override
    protected void onStart() {
        // 设置启用内置的缩放控件
        // 得到 mMapView 的控制权,可以用它控制和驱动平移和缩放
        MapController mMapController = mMapView.getController();
        // 获取传递过来的经纬度
        Intent intent = getIntent();
        String keys = intent.getStringExtra("key");
        String jingdu = intent.getStringExtra("jd");
        String weidu = intent.getStringExtra("wd");
        if (keys.equals("look")) {
            ausername = intent.getStringExtra("username");
            System.out.println("经纬度" + jingdu + "," + weidu);
            if (jingdu != null && jingdu != null) {
                // 经纬度
                List<String> list = new ArrayList<String>();
                list.add(ausername);
                list.add(weidu);
                list.add(jingdu);
                // 用给定的经纬度构造一个 GeoPoint,单位是微度 (度 * 1E6)
                GeoPoint point = new GeoPoint(
                    (int) (Double.valueOf(weidu) * 1E6),
                    (int) (Double.valueOf(jingdu) * 1E6));
                mMapController.setCenter(point);         // 设置地图中心点
                mMapController.setZoom(12);              // 设置地图 zoom 级别
                Log.d("mac", list.size() + "");
                if (list.size() > 0) {
                    upMapview(list);                     // 更新地图
                }
            }
        }
    }
```

```java
        // 给单选按钮组设置单击监听事件
        rg.setOnCheckedChangeListener(new OnCheckedChangeListener() {
            @Override
            public void onCheckedChanged(RadioGroup group, int checkedId) {
                // TODO Auto-generated method stub
                switch (checkedId) {
                // 选中普通地图
                case R.id.generalMap:
                    mMapView.setSatellite(false);
                    break;
                // 选中卫星地图
                case R.id.SatelliteMap:
                    mMapView.setSatellite(true);
                    break;
                default:
                    break;
                }
            }
        });
        super.onStart();
    }
}
```

（3）PosOverLay 类主要用于根据坐标位置在地图上绘制层或图标。首先重写 OverLay 的 draw() 方法，将定位坐标转换为屏幕坐标，然后使用画布类 Canvas 的对象方法绘制图标，重写 onTap() 方法在单击地图上的图标时弹出提示信息。部分关键程序如下：

```java
public PosOverLay(Drawable marker,
    Context context,List<String> list) {
    super(boundCenterBottom(marker));
    // 用给定的经纬度构造 GeoPoint，单位是微度（度 * 1E6）
    this.mContext = context;
    this.list=list;
    if(list.size()>0){
        GeoPoint p=null;
```

```
for(int i=0;i<list.size();i=i+3){
    p=new GeoPoint((int)(Double.valueOf
(list.get(i+1))*1E6),(int)(Double.valueOf
        (list.get(i+2))*1E6));
        GeoList.add(new OverlayItem(p,
        Integer.toString(i), list.get(i)));
    }
}
// createItem(int)方法构造 item。
//一旦有了数据, 在调用其他方法前, 首先调用这个方法
populate();
}
```

（4）AndroidManifest 配置。由于程序中需要使用 GPS 来获得定位信息所以要申明 GPS 的使用权限，地图是需要从网络上下载，所以要申明网络使用权限。

```
//申明 GPS 使用权限
<uses-permission android:name="android.permission.ACCESS_FINE_LOCATION">
</uses-permission>
<!-- Google Map 需要访问互联网, 所以必须授权 -->
<uses-permission android:name="android.permission.INTERNET" />
```

（5）客户端程序的打包和安装。直接利用 Eclipse 和 USB 连接线将项目运行在实验箱或手机中，即可完成客户端程序的安装。也可将项目下的 bin 目录中的 apk 文件拷贝到实验箱或手机的 sd 卡中进行安装。

4. 项目验证

（1）在图 19-7 所示系统主页中单击菜单，进入图 19-8 所示应用显示界面，单击"设置"进入设置界面，如图 19-9 所示，接着单击"无线和网络"进入网络设置界面，如图 19-10 所示。

图 19-7　主页

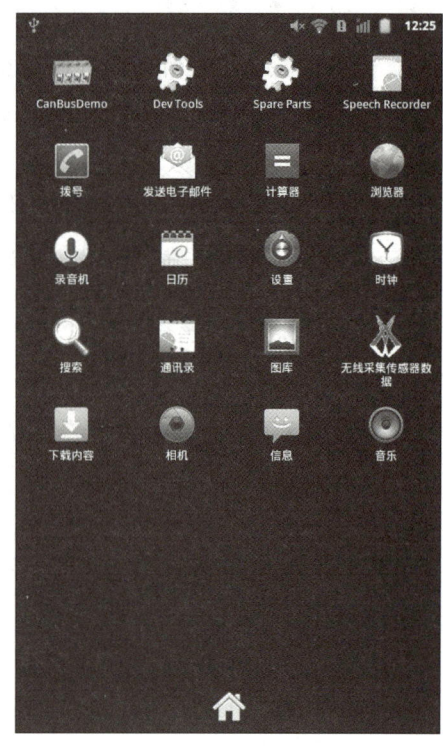
图 19-8　应用列表

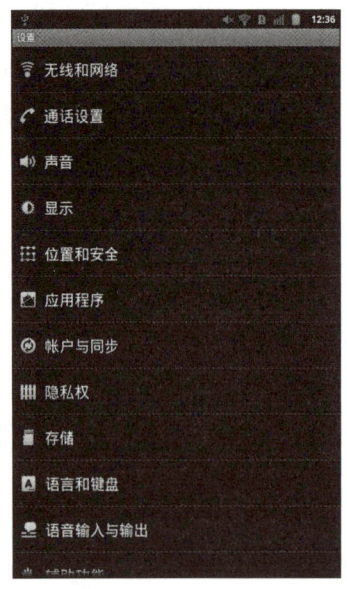
图 19-9　设置界面

图 19-10　网络设置界面

（2）单击"Wi-Fi"后的复选框按钮，打开 Wi-Fi。单击"Wi-Fi-设置"进入网络列表，如图 19-11 所示，单击网络列表中的网络名称，选择【连接】按钮，如图 19-12 所示，建立连接后在单击已连接的网络名称查看 IP 地址，如图 19-13 所示。

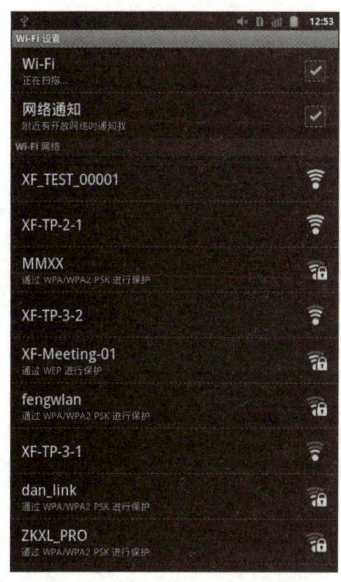 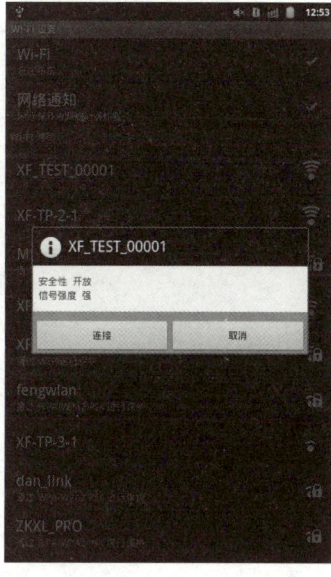 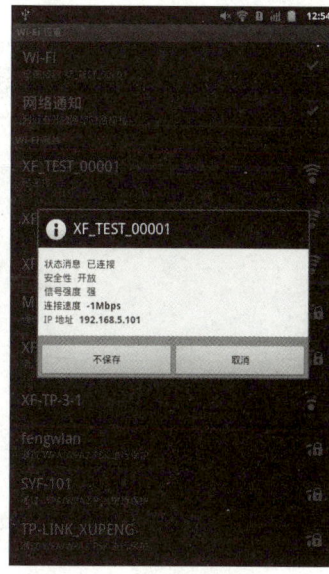

图 19-11　网络列表　　　　　图 19-12　建立连接　　　　　图 19-13　获取 IP 地址

（3）打开地图程序，查看当前的经纬度等信息，并显示在地图上，如图 19-14 和图 19-15 所示。

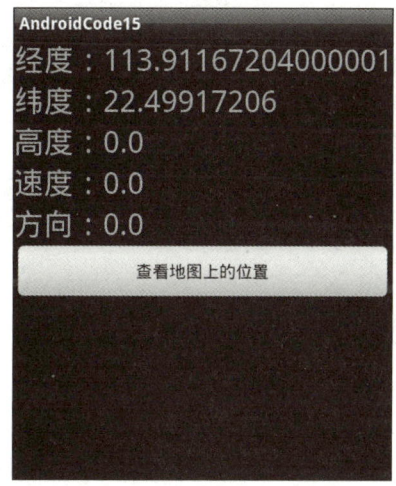 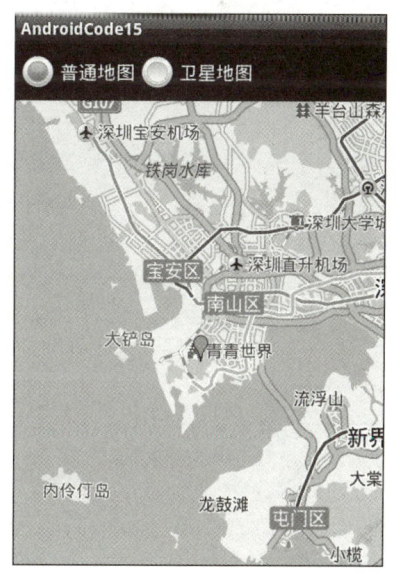

图 19-14　GPS 定位信息　　　　　　　图 19-15　当前位置

四、项目思考与扩展

1. 使用 LocationManager 怎么获得 GPS 的定位信息？
2. 使用 MapView 怎么实现自己的地图？

项目二十　桌面小组件

【本章导读】

　　Widget 是一个具有特定功能的视图，一般被嵌入到主屏幕中。用户在不启动任何程序的前提下，就可以在主屏幕上直接浏览 Widget 所显示的信息。本章项目所要讲的主要内容包括 Android 桌面小部件、App Widget 的开发入门指导，并通过一个简单实例的形式来直观的讲解 App Widget。

一、项目要求

1．掌握 Widget 的创建方法。
2．掌握后台 Service 更新 Widget 的方法。
3．掌握通过 Activity 与用户交互更新 Widget 的方法。

二、项目相关知识

　　应用程序窗口小部件（Widget）是微小的应用程序视图，可以被嵌入到其他应用程序中（比如桌面）并接收周期性的更新。你可以通过一个 App Widget provider 来发布一个 Widget。可以容纳其他 App Widget 的应用程序组件被称为 App Widget 宿主。
　　对于 widget 上，需要频繁更新的数据，一般都通过后台的 service 进行更新。当加载 widget 后，启动一个 service，并在 service 中开启一个线程更新 widget，销毁 widget 后，停止 service，同时停止更新线程。
　　当需要通过与用户交互达到更新 widget 的目的时，这时我们往往通过启动一个

activity，询问用户的需求，根据用户的要求更新 widget，并在更新后关闭 activity。

三、项目实施过程

下面我们就创建一个 widget。

1. 创建工程

（1）创建名为 AndroidCode20 的 Android 工程，包结构为"com.xdxy.widget"，如图 20-1 所示。Activity 名为 MainActivity，该 activity 用来与用户交互更新 widget 字体颜色。

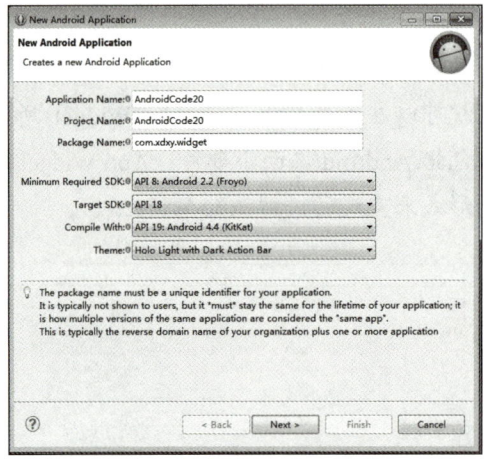

图 20-1　创建工程

（2）创建 TheDateofToday 类，该类继承自 AppWidgetProvider，用于更新 widget。创建 MainService 类，该类用于在后台更新时间时分秒，如图 20-2 所示。

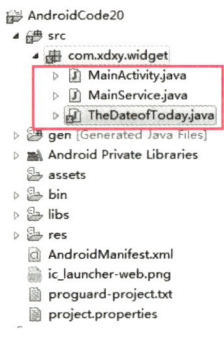

图 20-2　工程目录

2. XML 布局文件的开发

我们设计的目标界面如图 20-3 所示。

图 20-3　项目效果图

（1）在 res/layout 目录下新建 widget_layout.xml 文件定义 widget 布局。

```
<LinearLayout
    xmlns:android="http://schemas.android.com/apk/res/android"
    android:orientation="vertical"
    android:layout_width="wrap_content"
    android:layout_height="wrap_content"
    android:minHeight="146dp"
    android:minWidth="146dp"
    android:background="@drawable/bgblack"
    android:id="@+id/Base">
    <TextView
            android:layout_width="wrap_content"
            android:layout_height="wrap_content"
            android:id="@+id/Month"
```

```xml
            android:layout_gravity="center"
            android:textStyle="bold"
            android:textSize="18sp"
            android:textColor="#FFFFFF">
</TextView>
<TextView
            android:layout_width="wrap_content"
            android:layout_height="wrap_content"
            android:layout_gravity="center"
            android:id="@+id/Date"
            android:textStyle="bold"
            android:textSize="70sp"
            android:textColor="#FFFFFF">
</TextView>
<TextView
            android:layout_width="wrap_content"
            android:layout_height="wrap_content"
            android:id="@+id/WeekDay"
            android:layout_gravity="center"
            android:textSize="15sp"
            android:textStyle="bold"
            android:textColor="#FFFFFF">
</TextView>
<TextView
            android:layout_width="wrap_content"
            android:layout_height="wrap_content"
            android:id="@+id/time"
            android:layout_gravity="center"
            android:textSize="15sp"
            android:textStyle="bold"
            android:textColor="#FFFFFF">
</TextView>
</LinearLayout>
```

（2）在主配置文件中，进行注册。

```xml
<receiver android:label="@string/app_name" android:name=
                                      "com.xunfang.widget.TheDateofToday">
<intent-filter>
<action android:name="android.appwidget.action.APPWIDGET_UPDATE"></action>
<action android:name="com.xunfang.widget.updatetime"></action>
<action android:name="com.xunfang.widget.updateviews"></action>
</intent-filter>
<meta-data   android:name="android.appwidget.provider"
                             android:resource="@xml/widget"></meta-data>
</receiver>
```

（3）为 MainActivity 设置界面，界面文件为 main.xml，界面上设置 3 个按钮，用于设置 widget 字体颜色为红色、绿色和蓝色。

```xml
<LinearLayout xmlns:android="http://schemas.android.com/apk/res/android"
    android:layout_width="fill_parent"
    android:layout_height="fill_parent"
    android:orientation="vertical" >
    <LinearLayout
    android:layout_width="fill_parent"
    android:layout_height="300dp"
    android:orientation="vertical">

    <LinearLayout
    android:layout_width="fill_parent"
    android:layout_height="0dp"
    android:layout_weight="1.0">

    <TextView
    android:layout_width="fill_parent"
    android:layout_height="fill_parent"
    android:gravity="center_vertical|center_horizontal"
    android:textSize="40dp"
    android:text="更改字体颜色"/>
```

```xml
</LinearLayout>
<LinearLayout
android:layout_width="fill_parent"
android:layout_height="0dp"
android:layout_weight="2.0"
android:gravity="center_vertical|center_horizontal">

<Button
android:id="@+id/btFontToRed"
android:layout_width="wrap_content"
android:layout_height="wrap_content"
android:text="变红"
android:onClick="doClick"/>
<Button
android:id="@+id/btFontToGreen"
android:layout_marginLeft="70dp"
android:layout_width="wrap_content"
android:layout_height="wrap_content"
android:text="变绿"
android:onClick="doClick"/>
<Button
android:id="@+id/btFontToBlue"
android:layout_marginLeft="70dp"
android:layout_width="wrap_content"
android:layout_height="wrap_content"
android:text="变蓝"
android:onClick="doClick"/>
</LinearLayout>
</LinearLayout>
</LinearLayout>
```

3. Java 文件的开发

（1）在加载 widget 时，同时在后台启动服务；注销 widget 时，在后台销毁服务。

```java
public void onUpdate(Context context, AppWidgetManager appWidgetManager
        , int[] appWidgetIds){
    Intent intent = new Intent(context,MainService.class);
    context.startService(intent);
    RemoteViews updateView = buildUpdate(context);
    appWidgetManager.updateAppWidget(appWidgetIds, updateView);
    super.onUpdate(context, appWidgetManager, appWidgetIds);
}
@Override
public void onDeleted(Context context, int[] appWidgetIds) {
    // TODO Auto-generated method stub
    super.onDeleted(context, appWidgetIds);
    Intent intent = new Intent(context,MainService.class);
    context.stopService(intent);
}
```

(2)在 service 中,创建一个线程更新时间。

```java
new Thread(){
    public void run(){
        while(flag){
            long time=System.currentTimeMillis();
            long second=time/1000;
            second%=60;
            long minute=time/1000/60;
            minute%=60;
            long hour=time/1000/60/60;
            hour%=24;
            str=hour+"时"+minute+"分"+second+"秒";
            intent.putExtra("updatetime",str);
            sendBroadcast(intent);
            try {
                Thread.sleep(1000);
            } catch (InterruptedException e) {
                // TODO Auto-generated catch block
```

```
                    e.printStackTrace();
                }
            }
        }
    }.start();
}
```

（3）为 widget 添加单击事件，当单击时启动 activity。

```
private RemoteViews buildUpdate(Context context){
        RemoteViews updateView = null;
        Time time = new Time();
        time.setToNow();
        //第一行显示内容，月份和年份
        String month = months[time.month] + " " + time.year;
        //创建 RemoteViews 实例，设置布局
    updateView = new RemoteViews(context.getPackageName(), R.layout.widget_layout);
        //设置第一行内容
        updateView.setTextViewText(R.id.Month, month);
        //设置第二行内容
        updateView.setTextViewText(R.id.Date, new Integer(time.monthDay).toString());
        //设置第三行内容
        updateView.setTextViewText(R.id.WeekDay, days[time.weekDay]);
        Intent launchIntent = new Intent(context,MainActivity.class);
        launchIntent.setFlags(Intent.FLAG_ACTIVITY_NEW_TASK );
        PendingIntent intent = PendingIntent.getActivity(context, 0, launchIntent, 0);
        //为小控件指定当单击时响应的行为
        updateView.setOnClickPendingIntent(R.id.Base, intent);
        return updateView;
    }
```

（4）在 MainActivity 中为按钮添加单击事件。

```
public void doClick(View v){
        switch(v.getId()){
        case R.id.btFontToRed:
            intent.putExtra("updateviews",4);
```

```
                break;
            case R.id.btFontToGreen:
                intent.putExtra("updateviews",5);
                break;
            case R.id.btFontToBlue:
                intent.putExtra("updateviews",6);
                break;
        }
        sendBroadcast(intent);
        finish();
    }
```

（5）在 TheDateofToday 文件中，根据收到的广播，进行相应的更新。

```
switch(intent.getIntExtra("updateviews",0)){
        case 4:
            views.setTextColor(R.id.time,Color.RED);
            views.setTextColor(R.id.Month,Color.RED);
            views.setTextColor(R.id.Date,Color.RED);
            views.setTextColor(R.id.WeekDay,Color.RED);
            break;
        case 5:
            views.setTextColor(R.id.time,Color.GREEN);
            views.setTextColor(R.id.Month,Color.GREEN);
            views.setTextColor(R.id.Date,Color.GREEN);
            views.setTextColor(R.id.WeekDay,Color.GREEN);
            break;
        case 6:
            views.setTextColor(R.id.time,Color.BLUE);
            views.setTextColor(R.id.Month,Color.BLUE);
            views.setTextColor(R.id.Date,Color.BLUE);
            views.setTextColor(R.id.WeekDay,Color.BLUE);
            break;
    }
```

4. 项目验证

（1）将 Android 开发终端与 PC 机相连，并将本实验运行到 Android 开发终端上。

（2）在 Android 开发终端主界面长按，将呈现如图如图 20-4 所示界面，单击【widgets 项】。接着如图 20-5 所示，单击【今日日期】项。

 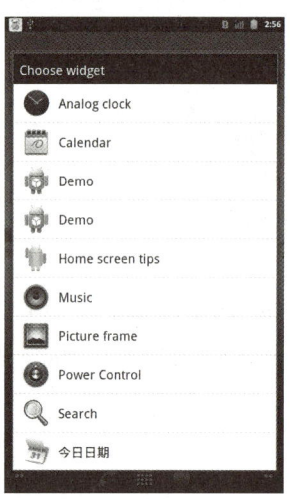

图 20-4　单击【widgets】项　　　　　图 20-5　单击【今日日期】项

（3）此时界面便会加载新建的 widget，widget 下部时间是实时变化的，如图 20-6 所示。单击 widget，则打开 activity，在此 activity 中可修改 widget 字体颜色，如图 20-7 所示。

图 20-6　widget　　　　　　　　　　图 20-7　activity 界面

（4）单击某个按钮，则 widget 中的字体便会变为相应的颜色，如图 20-8～图 20-10 所示。

图 20-8　变绿　　　　　　　图 20-9　变红　　　　　　　图 20-10　变蓝

四、项目思考与扩展

1. 为本实验中的 widget 添加修改日期功能。
2. 自己设计制作一个 widget。

项目二十一 NDK 的安装和使用

【本章导读】

　　Java 是半解释型语言，很容易被反汇编后拿到源代码文件。因此，在开发一些重要协议时，为了安全起见，通常使用 C 语言来编写，以增大系统的安全性。为此 Google 在 Android 方面为开发者提供了 NDK。本节，将学习在 Windows 下安装 NDK 环境！

一、项目要求

1. 了解 NDK 的作用。
2. 掌握 Cygwin 的安装和使用。
3. 掌握 NDK 的安装和使用。

二、项目相关知识

1. NDK（Native Development Kit）

　　NDK 是一系列工具的集合，帮助开发者快速开发 C（或 C++）的动态库。它集成了交叉编译器，并提供了相应的 mk 文件隔离 CPU、平台、ABI 等差异，开发人员只需要简单修改 mk 文件（指出"哪些文件需要编译""编译特性要求"等），就可以创建出 so。NDK 可以自动地将 so 和 Java 应用一起打包，极大地减轻了开发人员的打包工作。

　　NDK 提供了一份稳定、功能有限的 API 头文件声明。Google 明确声明该 API 是稳定的，在后续所有版本中都稳定支持当前发布的 API。从该版本的 NDK 中看出，这些 API 支持的功能非常有限，包含有：C 标准库（libc）、标准数学库（libm）、压缩库（libz）、Log 库（liblog）。

2．Cygwin

由于 NDK 编译代码时必须要用到 make 和 gcc，所以必须先搭建一个 Linux 环境，Cygwin 是一个在 Windows 平台上运行的 Unix 模拟环境，它对于学习 Unix/Linux 操作环境，或者从 Unix 到 Windows 的应用程序移植，非常有用。通过它，就可以在不安装 Linux 的情况下使用 NDK 来编译 C、C++代码了。

3．so 文件

我们通常把一些公用函数制作成函数库，供其他程序使用。函数库分为静态库和动态库两种。静态库在程序编译时会被连接到目标代码中，程序运行时将不再需要该静态库。动态库在程序编译时并不会被链接到目标代码中，而是在程序运行是才被载入，因此在程序运行时还需要动态库存在。

Linux 下的动态库的后缀是.so。so 文件是 Unix 的动态链接库，是二进制文件，作用相当于 Windows 下的.dll 文件。在 Android 中调用动态库文件（*.so）都是通过 jni 的方式。

三、项目实施过程

1．Cygwin

NDK 编译需要用到 Cygwin 中的 make 和 gcc，所以得先下载安装 Cygwin 软件。我们在光盘的实验目录中的 ch5 中的 Windows 目录下提供了安装文件。进入 www.cygwin.com 网站，单击 setup.exe 链接，下载 Cygwin 安装文件，如图 21-1 所示。

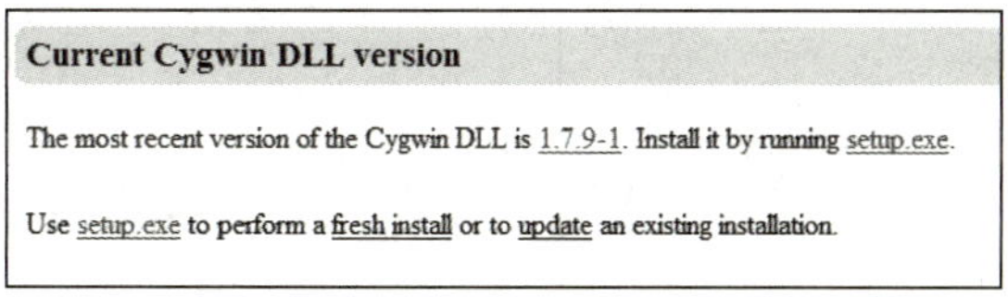

图 21-1　下载 Cygwin 安装文件

下载完成后，双击程序图标，进入安装界面，如图 21-2 所示。

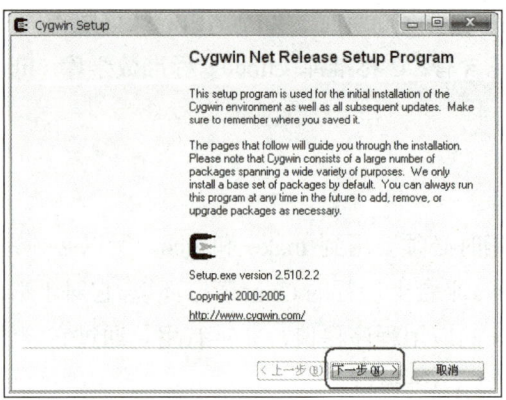

图 21-2　安装界面

安装过程比较简单，下面以图解的方式来说明，如图 21-3～图 21-8 所示。

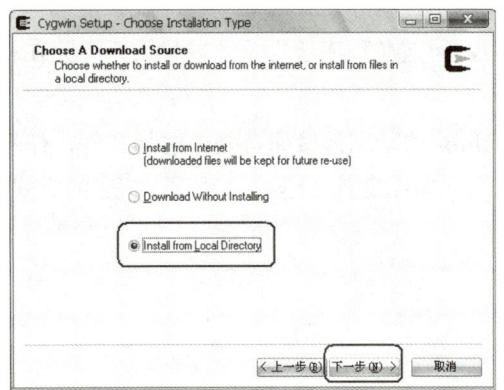

图 21-3　本地安装　　　　　　　　　　　图 21-4　安装的根目录（默认）

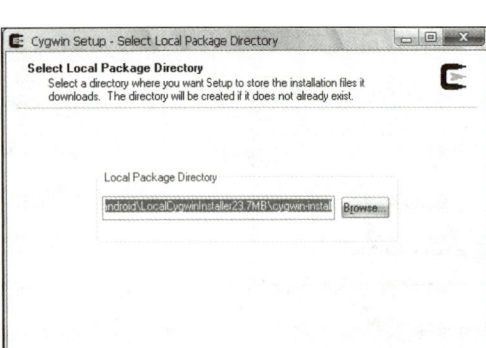
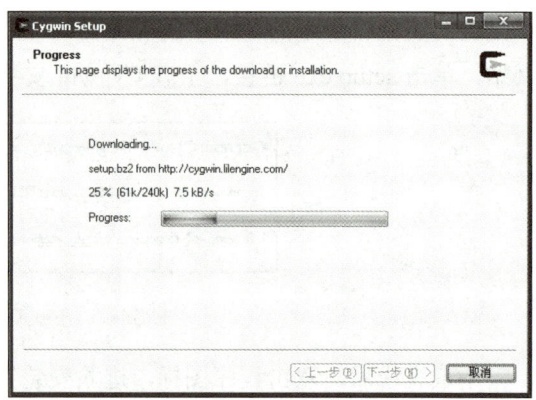

图 21-5　本地安装包的路径　　　　　　　图 21-6　下载文件

项目二十一　NDK 的安装和使用

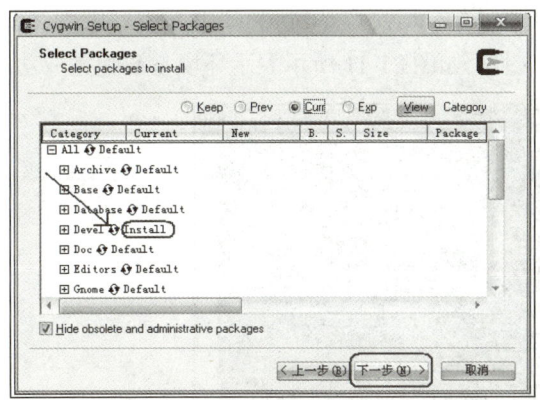

图 21-7　选择需要安装的包

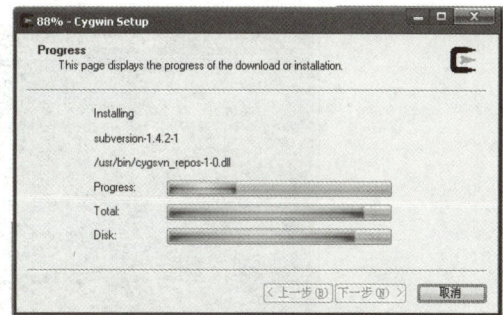

图 21-8　正在安装

NDK 需要的 make 和 gcc 在该节点下,单击红色箭头所指的圆圈箭头所指的地方,将 Default 状态切换成 Install 状态。

安装成功后,执行 Cygwin,进入图 21-9 所示界面。输入命令检查 make、gcc 是否安装成功,出现如图 21-10 所示信息,说明安装成功。

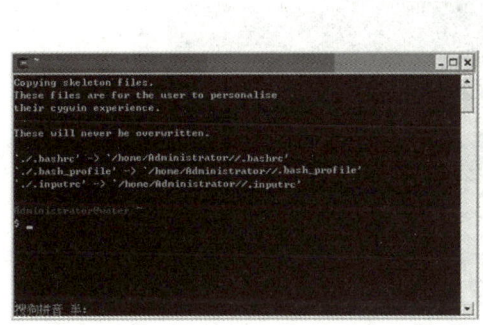

图 21-9　Cygwin 界面

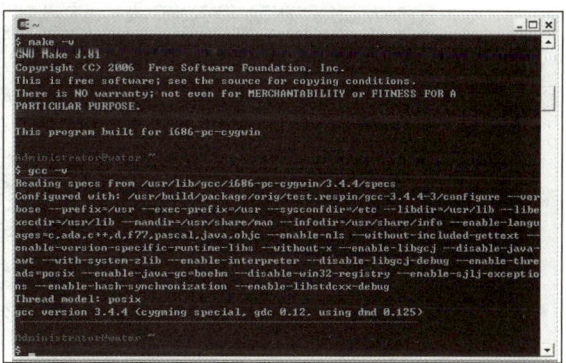

图 21-10　检查是否安装成功

2. NDK 环境参数的设置

可以从 http://developer.android.com/sdk/ndk/index.html 处下载最新的 Windows 下的 NDK 压缩包。

将下载的 NDK 压缩包解压到工作目录下,如笔者的工作路径为:D:android-ndk-1.6_r1。

配置 NDK，打开 Cygwin 软件，执行命令：cd /cygdrive/d/android-ndk-1.6_r1，进入存放 NDK 的路径下，执行./build/host-setup.sh 脚本文件，如图 21-11 所示表示环境参数配置成功。

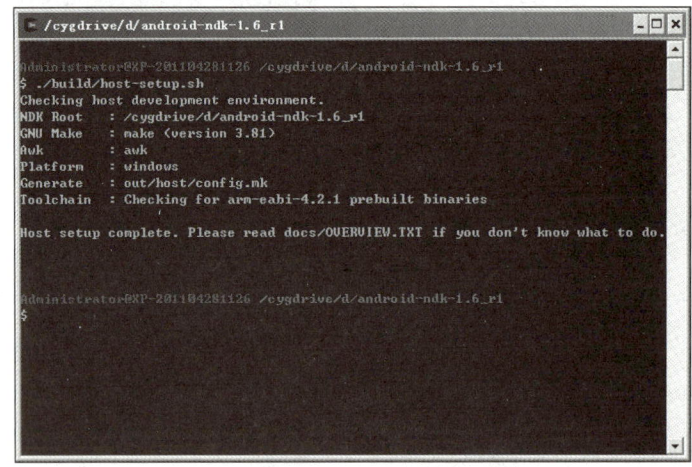

图 21-11　NDK 环境参数配置成功

到此，在 Windows 平台下成功搭建了 NDK 的开发环境。

3．Windows 环境下利用 NDK 生成 SO

启动 Cygwin，进入到 ndk 目录，执行 make APP=hello-jni 命令，如图 21-12 所示。

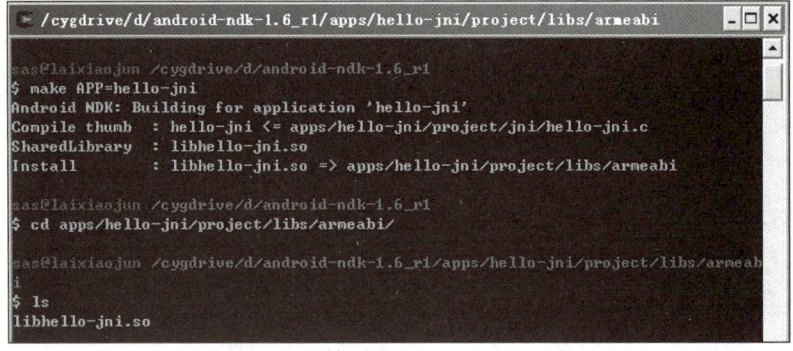

图 21-12　生成 SO

如图所示，生成了我们需要的 SO，这里需要说明的一点是 android 从 r3 以上的版本的 NDK 制作都需要 android 上层工程中的 AndroidManifest.xml 文件，没有的话就没有办法编译了。具体的上层如何进行 SO 的调用我们将下一个项目中进行介绍。

四、项目思考与扩展

1. 描述 Cygwin 的作用。
2. 描述 NDK 的作用。
3. 什么是 SO?

项目二十二 NDK 编译生成动态库

【本章导读】

NDK 的发布，使"Java+C"的开发方式终于转正，成为官方支持的开发方式。使用 NDK，我们可以将要求高性能的应用逻辑使用 C 开发，从而提高应用程序的执行效率。使用 NDK，我们可以将需要保密的应用逻辑使用 C 开发。通常使用 NDK 出于以下两个目的：

一是 Android 应用框架无法满足运行效率时；

二是需要使用大量的已有 C/C++源代码。

一、项目要求

1. 掌握动态库的生成过程。
2. 掌握动态库的调用过程。
3. 熟悉 Android.mk、Application.mk 文件的内容。

二、项目相关知识

当主机环境上部署 Java 平台，它可能或必须允许 Java 应用程序密切地与本机代码一起工作。程序员已经开始采用 Java 平台建立传统 C/C++语言编写的的应用程序。然而，由于现有遗留代码投资，Java 应用程序将在未来几年与 C 和 C++代码共存。JNI 是一个强大的功能，允许您利用 Java 的优势平台，但仍利用其他语言编写代码。作为一个 Java 虚拟机实现的一部分，JNI 是一个双向接口，以允许 Java 应用程序调用本地代码，反之亦然。

JNI 是 Java Native Interface 的缩写，中文为 JAVA 本地调用。从 Java1.1 开始，Java Native Interface（JNI）标准成为 Java 平台的一部分，它允许 Java 代码和其他语言写的代码进行交互。JNI 一开始是为了本地已编译语言，尤其是 C 和 C++而设计的，但是它并不妨碍你使用其他语言，只要调用约定受支持就可以了。

使用 Java 与本地已编译的代码交互，通常会丧失平台可移植性。但是，有些情况下 JNI 这样做是可以接受的，甚至是必需的，比如，使用一些旧的库，与硬件、操作系统进行交互，或者为了提高程序的性能。JNI 标准至少保证本地代码能工作在任何 Java 虚拟机上。

Android.mk 是 jni 根目录下必须存在描述 C/C++代码文件模块的文件，将代码模块的编译信息传递给 NDK 编译系统，是 NDK 编译系统编译脚本的一部分。

Application.mk 文件定义了应用程序编译的基本信息，是 Android NDK 编译系统中的非必备文件，如果出现，应保存在<Android NDK>/jni 目录中。

三、项目实施过程

1. 创建工程

打开 Eclipse，创建 Android 工程"AndroidCode22"如图 22-1 所示。

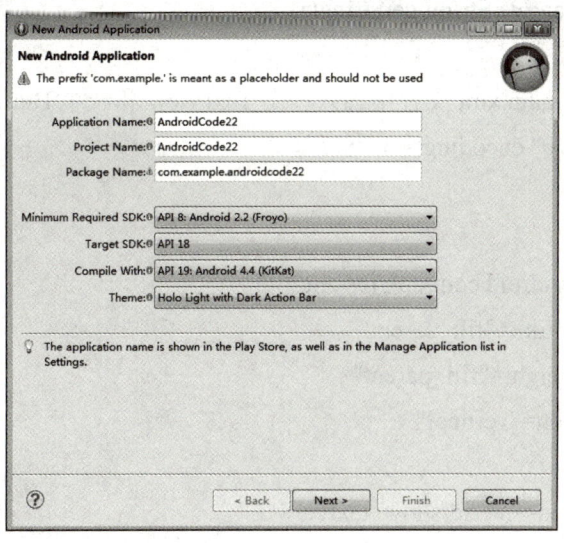

图 22-1 创建 Android 工程

2. Java 文件的开发

（1）在 Android 工程中创建 Java 类"Linux"与"HelloNDK"，在 Linux.Java 文件中输入如下所示内容。

```java
public class Linux {
    static {
        try {
            Log.i("JNI", "Trying to " + "load libled.so");
            /* 读取 libled.so 库 */
            System.loadLibrary("hndk");
        } catch (UnsatisfiedLinkError ule) {
            Log.e("JNI", "WARNING: " + "Could not load libhndk.so");
        }
    }
    /**
     * 获取字符串
     *
     * @return
     */
    public static native String getString();
}
```

（2）在 activity_main.xml 文件中加入一个 TextView 和一个 Button 如下所示。

```xml
<?xml version="1.0" encoding="utf-8"?>
<LinearLayout
  xmlns:android=
  "http://schemas.android.com/apk/res/android"
  android:layout_width="fill_parent"
  android:layout_height="fill_parent"
  android:orientation="vertical" >
<TextView
        android:id="@+id/show_textview"
        android:layout_width="fill_parent"
        android:layout_height="wrap_content"
        android:text="无内容"/>
```

```xml
<Button
    android:id="@+id/button_show"
    android:layout_width="wrap_content"
    android:layout_height="wrap_content"
    android:layout_gravity="center"
    android:text="获取字符串"/>
</LinearLayout>
```

(3) 在 HelloNDK.Java 中初始化控件，如下所示。

```java
public class HelloNDk extends Activity {
    private TextView show_textview;
    private Button button_show;
    @Override
    protected void onCreate(Bundle savedInstanceState) {
        super.onCreate(savedInstanceState);
        setContentView(R.layout.activity_main);
        show_textview = (TextView) findViewById(R.id.show_textview);
        button_show = (Button) findViewById(R.id.button_show);
    }
}
```

(4) 为按钮设置监听事件并调用 Linux.Java 中的方法，获取字符串并显示到 TextView 中，如下所示。

```java
button_show.setOnClickListener(new OnClickListener() {
    @Override
    public void onClick(View v) {
        // TODO Auto-generated method stub
        String message = Linux.getString();
        if (message != null) {
            show_textview.setText(message);
        } else {
            show_textview.setText("获取内容失败！");
        }
    }
});
```

（5）文件编译，编译 Java 文件成 C 语言的头文件。在开始功能表中单击执行，然后输入 cmd 进入 dos 命令窗口，如图 22-2 所示。

图 22-2　启动 dos 命令窗口

（6）进入到 android-ndk-1.6_r1 的 apps/目录下，如图 22-3 所示。

图 22-3　进入 NDK 目录

（7）在 apps 目录下有系统已经存在的工程文件，同样，我们也在该目录下创建我们的工程文件 hndk。依次执行下列指令，如图 22-4 所示。

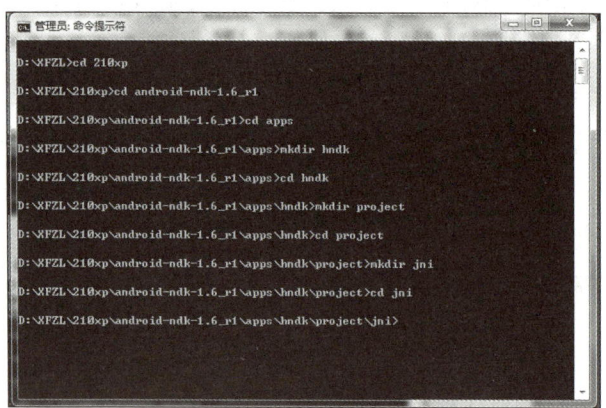

图 22-4　创建 project 等目录

mkdir hndk：创建一个文件夹，Led 为创建的 NDK 工程名。

mkdir project：创建一个文件夹，此文件夹下将存放 C 及 C++原始文件案，编译后生成的\libs\armeabi\libhndk.so 文件也在此目录下。

> mkdir jni：创建一个用于存放 C 及 C++原始程序码文件，以及要编译这些程序码的脚本文件。

（8）执行命令：Javah–classpath "C:\Users\Administrator\workspace\AndroidCode18\bin\classes" com.xunfang.ndk.Linux，如图 22-5 所示，在当前目录 jni 下生成了一个 com_xunfang_ndk_Linux.h 文件。

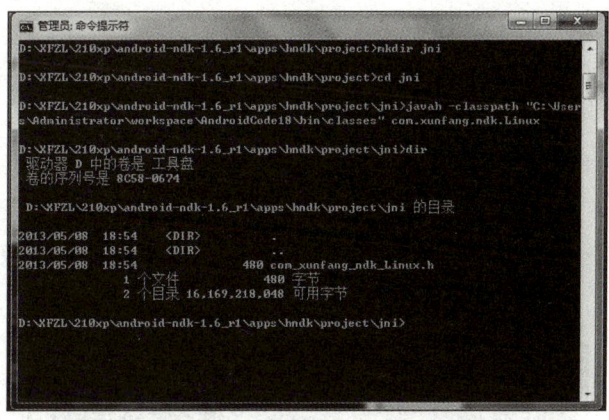

图 22-5　查看 jni 目录中的文件

（9）在 jni 目录下创建 com_xunfang_ndk_Linux.c 文件，编写调用 C 语言头的 C 文件。

#include <jni.h>
#include "com_xunfang_ndk_Linux.h"
#include <stdio.h>
#include <stdlib.h>
#include <fcntl.h>
#include <errno.h>
#include <unistd.h>
JNIEXPORT jString JNICALL Java_com_xunfang_ndk_Linux_getString
(JNIEnv *env, jclass mc){
　jstring jstr;
　char cstr[]="Hello NDK";
　jstr=env->NewStringUTF(cstr);
　return jstr;
}

（10）SO 配置文件生成。

现在 jni 目录下已经有 com_xunfang_ndk_Linux.c、com_xunfang_ndk_Linux.h 这两个文件。还需要添加一个脚本文件实现编译连接，其文件名为 Android.mk，其内容如下：

LOCAL_PATH := $(call my-dir)
include $(CLEAR_VARS)
LOCAL_MODULE := hndk
LOCAL_SRC_FILES := com_xunfang_ndk_Linux.c
include $(BUILD_SHARED_LIBRARY)

以上步骤完成后，进入到\apps\led 目录下，此时需要为 project 添加一个编译脚本 Application.mk，内容如下：

APP_PROJECT_PATH := $(call my-dir)/project
APP_MODULES := hndk

（11）SO 动态链接库生成，打开 Cygwin 终端，进入到 ndk 目录，然后执行命令：make APP=hndk 即可，如图 22-6 所示。

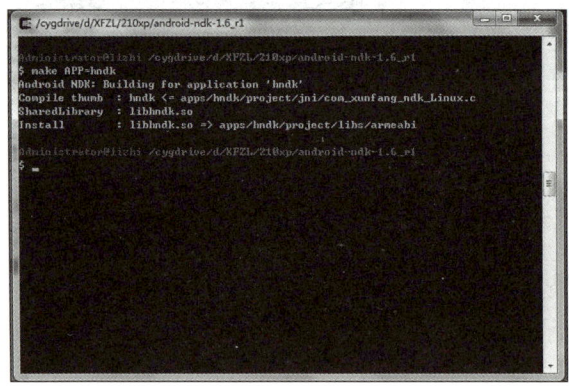

图 22-6　生成动态链接库

（12）添加 SO 库，我们在上面已经编译出动态的链接库，现在我们找到这个链接库，如图 22-7 所示。

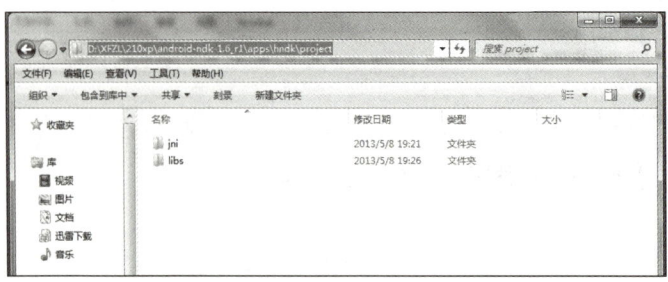

图 22-7　拷贝 libs 文件夹

项目二十二 NDK 编译生成动态库

（13）把这个文件夹下面的 libs 文件夹拷到你的工程目录下，如图 22-8 所示。

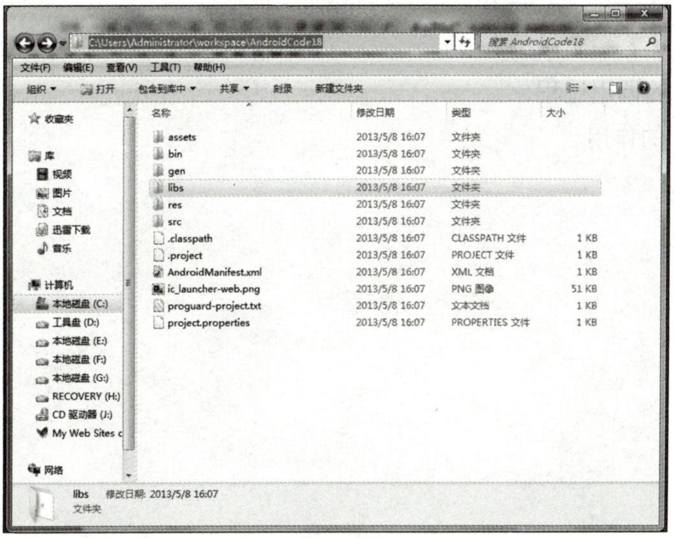

图 22-8 把 libs 目录拷贝到 Led 应用程序目录

（14）生成 APK 由于 eclipse 可以设定自动编译功能，所以当你把 SO 拷贝到工程下时，它也已经编译完成了，生成的 APK 在工程目录下的 bin 文件夹下，如图 22-9 所示。

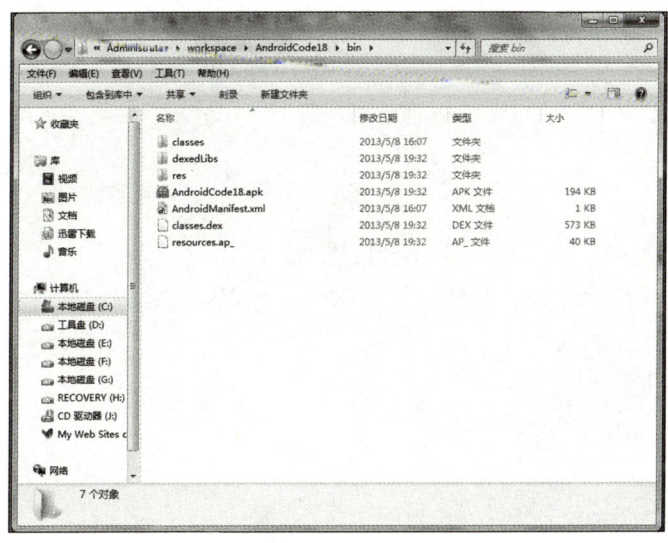

图 22-9 生成的 APK

四、项目思考与扩展

1. 描述 SO 的调用过程。
2. 描述 JNI 的作用。

参考文献

［1］靳岩．Android 开发入门与实战［M］．北京：人民邮电出版社，2011．

［2］李兴华．名师讲坛——Android 开发实战经典［M］．北京：清华大学出版社，2012．

［3］邓凡平．深入理解 Android：卷Ⅰ［M］．北京：机械工业出版社，2011．

［4］李刚．疯狂 Android 讲义（第 2 版）［M］．北京：电子工业出版社，2013．

［5］李宁．Android 开发权威指南［M］．北京：人民邮电出版社，2011．

［6］杨云君．Android 的设计与实现：卷Ⅰ［M］．北京：机械工业出版社，2013．